Just-in-Time
Algebra and Trigonometry

FOR STUDENTS OF CALCULUS

Guntram Mueller
Ronald I. Brent
University of Massachusetts-Lowell

ADDISON-WESLEY

An imprint of Addison Wesley Longman, Inc.

Reading, Massachusetts • Menlo Park, California • New York • Harlow, England
Don Mills, Ontario • Sydney • Mexico City • Madrid • Amsterdam

Cover Illustration by Toni St. Regis.

Reproduced by Addison-Wesley from camera-ready copy supplied by the authors.

ISBN   0-201-41951-3

1 2 3 4 5 6 7 8 9 10 CRS 99989796

Table of Contents

To the Student

Okay, you're in college, you're taking calculus, and as you flip through the pages of your calculus text, you see a lot of things that look suspiciously like algebra and trig. In fact, you see algebra on every page, along with some other odd characters like $\int$ and Δ and ∇. What's more, your professor lectures about the importance of algebra and trig in calculus, how they are the language of calculus, and you begin to pay a little more attention as you reflect on your own algebra and trig experience. You wonder: Is my background strong enough to handle calculus? Did they teach me the right things in high school? And did I learn them? Should I have spent more time on homework? (Ugh!) Will it all come back now to haunt me? (I need some air !) Where can I look up this stuff ? And what do I need to look up anyway?

If this strikes a chord with you, take a close look at this book. It contains algebra and trig, and it is arranged in just the order that you'll need it, exactly at the moment when it comes up in your calculus course. That's the idea of *Just-in-Time Algebra and Trigonometry*.

For example, take a look at the table of contents. When you take equations of circles and other conic sections in calculus, you need to know how to complete the square from algebra. There it is! When you take limits in calculus, you need to know how to factor from algebra. It's right there, just when you need it. No chasing around, no wading through piles of stuff you don't need. Just what you need, just when you need it, to allow you to calculate the limits. When you use implicit differentiation, and need to solve some strange-looking equations of degree 1, here they are, to allow you to navigate with confidence and accuracy. Obviously, accuracy is important. Even profound ideas, without accuracy, can make bridges fall, buildings collapse, and rockets blow up. Not to mention how your boss might react!

What is the best way to use this book? Here's how. At any point in your calculus course:

a) Read the corresponding part of this book as directed by the table of contents, having pencil and paper at hand. If there are things you don't understand, or that just seem wrong (it happens!), put a question mark in the book and ask your professor in class. Make sure it all makes sense <u>to you</u>. <u>Your mind is the ultimate judge!</u>

b) It's best if you do <u>all</u> of the exercises. The exercise sets are kept deliberately short, and each exercise brings out one or more new features.

Mathematics is not a subject like other subjects. It may not always be an easy subject to learn, and you should not be discouraged if it does not come right away, but with hard work, determination, and care, you <u>can</u> succeed in calculus. Mathematics is unique unto itself in its representation of scientific thought. It is the language of science and technology. Mathematics is not mysterious, it can be learned, and for scientists and engineers it is absolutely essential.

To the Instructor

What calculus instructors have not, at one time or another, wrung their hands or pulled out their hair in despair over the level of competence of their students? This book is an attempt to try to salvage whatever is left of your hair.

As you can see from the table of contents, this book is organized by topic from most standard calculus courses. When limits are to be calculated, for example, the student needs factoring skills, so here are the factoring methods, all lined up at just the time the student needs them. There is too little room in calculus books for this material on factoring, and sending students to the library to look up factoring methods on their own does not produce the desired results.

Similarly, when studying conics, here's the method of completing the square. When studying the idea of the derivative, here is a review of rational operations on rational expressions. Doing implicit differentiation? Here's how to solve equations of degree 1, even strange-looking ones. The idea is that the student does not need to hunt through libraries and unfamiliar algebra texts, to look for what he or she might need. Rather, it is all lined up here, "just-in-time," exactly when needed.

What is not in this book? This book does not contain material that is presented in standard calculus texts. So, for example, limits are not included, nor is there any intensive treatment of inverse trig functions, since they are covered in the standard texts. We have intentionally left out material on graphing calculators since this is already covered in calculus books or done explicitly in calculus class. The idea is to keep this book light enough for students to actually carry it and use it – and cheap enough for them to afford it.

This book is intended to be used in one of two ways:

a) As a second text for calculus courses whose syllabi officially include a review of algebra and trigonometry. The organization of this book is a key feature: It can be used with almost all standard calculus texts, including the reform texts, in a manner that interleaves the topics from algebra and trigonometry with those from calculus. By timing certain topics in algebra and trigonometry to be given just before, or while, they are needed in calculus, we address the often-heard student complaint of irrelevance. For example, factoring is made more pertinent by tying it in with limit problems that can't be done except by factoring. This will enhance the motivation to learn factoring, a topic that can well use all the motivation it can get.

b) As a companion to a standard calculus I course, for the majority of students, who in fact need a little well-timed help in algebra and trigonometry. This book is written in an easy style, that can be understood by all students on their own, without input by the instructor. It is there as a reference, a guide, a handbook, a companion on a journey that is rewarding, but where the traveler could sometimes use a little support.

The idea for both uses is the same: the deft timing of the algebra and trigonometry review, topic-by-topic, just at the time that they are needed in calculus. It makes for a more relevant presentation of the review topic, which allows the student to be more interested in it because he or she is about to have to use it in calculus homework, that night or the next.

Acknowledgments

We wish to acknowledge all those who helped in this venture, from our wives Edie and Leor and our children Ariadne and Sarah, to the many students who have helped test the book. We especially wish to thank Joyce Williams and Jacob Weinberg, our colleagues at University of Massachusetts-Lowell, Laurie Rosatone and Jennifer Wall at Addison Wesley Longman, and all the people who reviewed the manuscript. They provided us with excellent insights and many valuable suggestions that really brought the final copy together. Lastly, we thank Jerry Garcia for all the years of excellent music and vibes.

Once in a while you get shown the light, in the strangest of places, if you look at it right.

– Robert Hunter

University of Massachusetts-Lowell   Guntram Mueller

Ronald I. Brent

Chapter 1

Numbers and Their Disguises

Every number can be written in many different forms. For example, the numbers $\frac{8}{12}$, $\frac{10}{15}$, $\dfrac{3}{4 + \pi/2\pi}$, and even $\dfrac{|4x| + 4}{|6x| + 6}$, are all really just different ways of writing the number $\frac{2}{3}$. (Check it out; don't take <u>our</u> word for it.) For most purposes, the idea is to keep things as simple as possible, and just to use the form $\frac{2}{3}$.

Sometimes it is better to have a mathematical expression written as a product, other times it is better to have a sum; it all depends on what you need to do with it. In any event, <u>changing the form of an expression is something you have to do all the time</u>. Correctly! It's the nuts and bolts of mathematics, and is used in all the sciences, engineering, and even economics and medicine.

1.1 <u>Brackets</u>

Brackets are ways of "packaging" or grouping numbers together.

Example 1:

$$5 - (1 + \frac{1}{2}) + (3 - 4) - (7 - \frac{1}{2})$$

$$= 5 - (1\tfrac{1}{2}) + (-1) - (6\tfrac{1}{2})$$

$$= 5 - \frac{3}{2} - 1 - \frac{13}{2}$$

$$= 5 - 1 - \frac{3}{2} - \frac{13}{2}$$

$$= 4 - \frac{16}{2} \;=\; 4 - 8 \;=\; -4 \quad \blacksquare$$

You may prefer to get rid of the brackets. Here's how: a) if there's a "+" in front, leave all the signs of the terms inside as they are; b) if there is a "−" in front, change all the signs of the terms inside.

Example 2:

$$5 - \left(1 + \frac{1}{2}\right) + (3 - 4) - \left(7 - \frac{1}{2}\right)$$

$$= 5 - 1 - \frac{1}{2} + 3 - 4 - 7 + \frac{1}{2}$$

$$= 5 - 1 + 3 - 4 - 7 - \frac{1}{2} + \frac{1}{2}$$

$$= -4 \quad \blacksquare$$

Example 3:   Simplify $3 - 2(8 + 1) - 3(5 - 7)$

Solution: Method 1: $3 - 2(9) - 3(-2) = 3 - 18 + 6 = -9$

Method 2: $3 - 16 - 2 - 15 + 21 = -9 \quad \blacksquare$

Notice that in the second method in Example 3, the 8 and 1 were both multiplied by -2, and the 5 and -7 were multiplied by -3. This is based on the law of distributivity – namely, $a(b + c) = ab + ac$. Method 1 is easier in this case, but in many other cases method 2 will be needed. You've got to know both!

In algebra, letters stand for numbers, so the laws of arithmetic apply to them in exactly the same way:

Example 4:

$$xy - (2x - y) - 2y(1 - x)$$

$$= xy - 2x + y - 2y + 2yx$$

$$= 3xy - 2x - y \quad \blacksquare$$

Notice that xy and $2yx$ add up to $3xy$.

Example 5:

$$3x^2y - (x^2 - y^3) - 2y(x - y)$$

$$= 3x^2y - x^2 + y^3 - 2xy + 2y^2 \quad \blacksquare$$

Notice the "+" in front of the y in Example 4 and the y^3 in Example 5, as well as the use of the distributive law in both examples. If the exponents in the last example are confusing you, skip ahead to section 1.4 and then return here.

Exercises 1.1 Simplify:

1) $4 - (5 - 3) + 3(4 - 7) - 4(1 + 2)$

2) $-2(3 - 4) + 4(5 + 3) - 3(2 - 6)$

3) $4xy - (x - 2xy) - 2y(x - 1)$

4) $(s - t) - (u - t) - (v - u) - (s - v)$

5) $2x(y - 3) - y(x + xy) + 2y(x + 1)$

6) $x(y + z) - z(x + y) + 2y(x - z) - x(3y - 2z)$

7) $xy(x + y^2) - (2x^2y^2 - 2xy^3) - 2y^2(x^2 - y + xy)$

8) $xy^2(x^2 + y^2) - 3x(2xy^2 - 2xy^3) + 2y^2(x - xy^2 - x^2)$

9)   Consider simplifying the expression $-2(7 - 4) + 5(2 + 3) - 2(2 - 6)$ by first computing the expression within the brackets and then by using the distributive law. Keep track of the total number of computations (additions and subtractions or multiplications) in each method. Which way is "cheaper" ?

1.2 <u>Multiplying and Dividing Fractions</u> (We'll do adding and subtracting later.)

Multiplying fractions is the easiest manipulative task. The numerator is the product of the given numerators, and the denominator is the product of the given denominators. That is

$$\frac{a}{b} \cdot \frac{c}{d} = \frac{a \cdot c}{b \cdot d} .$$

We use the symbol "·" to denote multiplication, although sometimes we will omit the symbol altogether.

Example 1:

a) $\dfrac{2}{3} \cdot \dfrac{5}{7} = \dfrac{2 \cdot 5}{3 \cdot 7} = \dfrac{10}{21}$

b) $\dfrac{1}{9} \cdot \left(-\dfrac{5}{8} \right) = \dfrac{1}{9} \cdot \dfrac{-5}{8} = \dfrac{1 \cdot (-5)}{9 \cdot 8} = \dfrac{-5}{72}$ ∎

This rule can be extended to multiplying more than two fractions simply by multiplying across all the numerators and denominators.

Example 2:

a) $\dfrac{1}{4} \cdot \dfrac{7}{5} \cdot \dfrac{3}{8} = \dfrac{1 \cdot 7 \cdot 3}{4 \cdot 5 \cdot 8} = \dfrac{21}{160}$

b) $\dfrac{-1}{7} \cdot \dfrac{3}{8} \cdot \dfrac{-2}{\pi} = \dfrac{(-1)(3)(-2)}{7 \cdot 8 \cdot \pi} = \dfrac{6}{56\pi} = \dfrac{3}{28\pi}$ ∎

Of course, multiplying a number by 1 produces the same number, so: $\dfrac{5}{6} \cdot 1 = \dfrac{5}{6} \cdot \dfrac{4}{4} = \dfrac{20}{24}$. Going backward is usually more important: $\dfrac{20}{24} = \dfrac{5 \cdot 4}{6 \cdot 4} = \dfrac{5 \cdot \cancel{4}}{6 \cdot \cancel{4}} = \dfrac{5}{6}$. Here's the key point: you can cancel the factor 4 in the numerator and the denominator. More generally, if a number $c \neq 0$ is a factor of both the top and bottom of a fraction, it may be canceled. When all those common factors are canceled, the fraction is said to be in <u>lowest terms</u>.

Example 3:   Put $\dfrac{30}{84}$ into lowest terms.

Solution: First factor the numerator and denominator as much as possible and then cancel all common factors.

$$\frac{30}{84} = \frac{\cancel{2}\cdot\cancel{3}\cdot 5}{2\cdot\cancel{2}\cdot\cancel{3}\cdot 7} = \frac{5}{2\cdot 7} = \frac{5}{14} \quad \blacksquare$$

Warning: <u>Make sure you cancel only those numbers that are factors of the entire top and the entire bottom.</u> Don't be tempted to try "creative canceling." Consider the expression

$$\frac{3(8) + 7(5)}{3(4)}.$$

Can you cancel the 3's ? NO NO NO NO NO NO NO NO NO !!!! Get the picture? The problem is that 3 is not a factor of the <u>entire</u> numerator, only of its first term. Hence you cannot simply cancel the 3's. If the expression had been

$$\frac{3(8) + 3(5)}{3(4)},$$

then you would be able to cancel the 3's. In this case, <u>each term of the numerator</u> contains the factor of 3 that is also in the denominator.

$$\frac{3(8) + 3(5)}{3(4)} = \frac{3(8 + 5)}{3(4)} = \frac{\cancel{3}}{\cancel{3}}\cdot\frac{(8 + 5)}{(4)} = 1\cdot\frac{(8 + 5)}{(4)} = \frac{13}{4}.$$

Dividing by a fraction is done by inverting <u>that</u> fraction and multiplying:

$$\frac{\dfrac{a}{b}}{\dfrac{c}{d}} = \frac{a}{b}\cdot\frac{d}{c} = \frac{a\cdot d}{b\cdot c}.$$

For example,

$$\frac{\dfrac{-1}{3}}{\dfrac{5}{6}} = \frac{-1}{3}\cdot\frac{6}{5} = \frac{-6}{15} = \frac{-2}{5},$$

or you can cancel early: $\dfrac{-1}{1\not{7}}\cdot\dfrac{\not{6}2}{5}=\dfrac{-2}{5}$

Example 4: Simplify the following expressions:

a) $\dfrac{\frac{2}{3}}{\frac{3}{8}}$ b) $\dfrac{\frac{5}{4}}{\frac{-10}{3}}$

Solution:

a) $\dfrac{\frac{2}{3}}{\frac{3}{8}} = \dfrac{2}{3}\cdot\dfrac{8}{3} = \dfrac{16}{9}$

b) $\dfrac{\frac{5}{4}}{\frac{-10}{3}} = \dfrac{5}{4}\cdot\dfrac{3}{-10} = -\dfrac{3}{8}$ ∎

Example 5: Simplify the expression $\left(\dfrac{x+1}{y}\right)\cdot\left(\dfrac{-y+1}{x}\right)$.

Solution: $\left(\dfrac{x+1}{y}\right)\cdot\left(\dfrac{-y+1}{x}\right) = \dfrac{(x+1)\cdot(-y+1)}{y\cdot x}$

$= \dfrac{-xy+x-y+1}{xy}$

$= \dfrac{1-xy+x-y}{xy}$ ∎

Example 6: Simplify the expression $\dfrac{\frac{x^2 y}{z}}{\frac{xy^2}{z^3}}$.

Solution: $\dfrac{\frac{x^2 y}{z}}{\frac{xy^2}{z^3}} = \dfrac{x^2 y}{z}\cdot\dfrac{z^3}{xy^2}$

$= \dfrac{x^2 y z^3}{z x y^2} = \dfrac{x z^2}{y}$ ∎

Again, if you are confused by the exponents, skip ahead to section 1.4 and then return here.

In calculus, we use numbers called real numbers. These can be thought of as all decimal numbers, including those having an infinite number of digits. The set of real numbers corresponds to the set of points on the number axis. Rational numbers are those real numbers that can be expressed as a quotient of two integers. For example, the numbers $\frac{1}{3}$, $\frac{4}{10}$, -5, $\frac{18}{7}$, $\sqrt{64}$, and 3.857 are all rational numbers. (Convince yourself of this by expressing them as the quotient of two integers.) All other real numbers are called irrational. The number $\sqrt{2}$ can be easily shown to be irrational. The number π is also irrational but that's much harder to prove.

Exercises 1.2 Multiply out and reduce to lowest terms. (Some answers may have more than one form.)

1) $\dfrac{5}{16} \cdot \dfrac{8}{10}$

2) $\dfrac{2\pi}{3} \cdot \dfrac{3\pi}{4}$

3) $\dfrac{-1}{3} \cdot \dfrac{-9}{5}$

4) $\dfrac{\frac{4}{75}}{\frac{8}{25}}$

5) $\dfrac{\frac{-7}{51}}{\frac{3}{12}}$

6) $\dfrac{\frac{3\pi}{7}}{\frac{2\pi}{3}}$

7) $\dfrac{7x}{3y} \cdot \dfrac{3y+2}{x}$

8) $\left(\dfrac{x+2}{1+y}\right) \cdot \left(\dfrac{1-y}{x}\right)$

9) $\dfrac{\frac{xy}{w}}{\frac{xy-2x}{w}}$

10) $\dfrac{xy}{wz} \cdot \dfrac{w^2 z}{x^2 y^2}$

11) $\dfrac{\frac{xy}{(x+y)}}{\frac{x^2 y}{(x+y)^3}}$

12) $\dfrac{\frac{xy}{(x-y)}}{\frac{x^2}{y} \cdot \frac{y^3}{x}}$

1.3 Adding and Subtracting Fractions

Adding and subtracting fractions is easy when the denominators are all the same. For example,

$$\frac{2}{5} + \frac{17}{5} - \frac{4}{5} = \frac{2 + 17 - 4}{5} = \frac{15}{5} = 3.$$

But what if the fractions don't all have the same denominator? Then you first need to rewrite the fractions so that they all have the same denominator, called a <u>common denominator</u>. For example, if you want to add $\frac{2}{3} + \frac{4}{5}$, you can use a common denominator of 15. So:

$$\frac{2}{3} + \frac{4}{5} = \frac{10}{15} + \frac{12}{15} = \frac{22}{15}.$$

Remember, to get a common denominator you can always use the product of the individual denominators ($3 \cdot 5 = 15$), but sometimes even a smaller number will do it. Then, to get the new numerator,

$$\frac{2}{3} = \frac{?}{15},$$

divide 3 into 15, to get 5, then multiply by 2, to get 10. Try it: $\frac{3}{7} = \frac{?}{42}$. (We're doing the opposite of canceling common factors.)

Example 1:   a)   Simplify $\frac{3}{5} + \frac{1}{2} - \frac{2}{3}$.

Solution: $5 \cdot 2 \cdot 3 = 30$ will serve as a common denominator. So:

$$\frac{3}{5} + \frac{1}{2} - \frac{2}{3} = \frac{18}{30} + \frac{15}{30} - \frac{20}{30}$$

$$= \frac{18 + 15 - 20}{30} = \frac{13}{30}$$

b)   Simplify $\frac{1}{6} - \frac{1}{9}$.

Solution: We could use $6 \cdot 9 = 54$ as a common denominator, but even 18 will do nicely, because <u>both 6 and 9 divide evenly into 18</u>. So:

$$\frac{1}{6} - \frac{1}{9} = \frac{3}{18} - \frac{2}{18} = \frac{1}{18}$$

c)   Simplify $\dfrac{\dfrac{1}{2} + \dfrac{3}{4}}{\dfrac{1}{3} - \dfrac{1}{6}}$.

Solution: $\dfrac{\dfrac{1}{2} + \dfrac{3}{4}}{\dfrac{1}{3} - \dfrac{1}{6}} = \dfrac{\dfrac{2+3}{4}}{\dfrac{2-1}{6}} = \dfrac{\dfrac{5}{4}}{\dfrac{1}{6}} = \dfrac{5}{24} \cdot \dfrac{\cancel{6}3}{1} = \dfrac{15}{2}$ ∎

You will also have to be comfortable with adding and subtracting fractional expressions involving variables.

Example 2:

$$\frac{xy}{z} + \frac{x}{5} = \frac{5xy}{5z} + \frac{xz}{5z} = \frac{5xy + xz}{5z}$$ ∎

Example 3:

$$\frac{\dfrac{3a+2b}{5ab}}{\dfrac{a}{b} - \dfrac{b}{a}} = \frac{\dfrac{3a+2b}{5ab}}{\dfrac{a^2 - b^2}{ab}}$$

$$= \frac{3a+2b}{5ab} \cdot \frac{ab}{a^2 - b^2}$$

$$= \frac{3a+2b}{5a^2 - 5b^2}$$ ∎

Exercises 1.3 Express as a single fraction and simplify:

1) $\dfrac{1}{3} + \dfrac{1}{4}$

2) $\dfrac{7}{6} + \dfrac{5}{24}$

3) $\dfrac{2}{5} - \dfrac{1}{2} + \dfrac{1}{3}$

4) $\dfrac{1}{2} - \dfrac{1}{4} + \dfrac{1}{8} - \dfrac{1}{16}$

5) $\dfrac{1}{2} + \dfrac{4}{3} - \dfrac{2}{5} - \dfrac{3}{15}$

6) $\dfrac{5}{6} - \dfrac{4}{3} + \dfrac{2}{9} - \dfrac{3}{2}$

7) $\dfrac{\dfrac{1}{3} + \dfrac{2}{5}}{\dfrac{3}{2}}$

8) $\dfrac{\dfrac{1}{4} - \dfrac{2}{3}}{\dfrac{3}{2} - \dfrac{2}{5}}$

9) $\dfrac{\dfrac{2}{7} + \dfrac{1}{3}}{\dfrac{4}{3} + \dfrac{2}{5}} + \dfrac{1}{3}$

10) $\dfrac{1}{x} + \dfrac{1}{y}$

11) $\dfrac{1}{y} - \dfrac{1}{x}$

12) $\dfrac{4}{x} - \dfrac{2}{y} + \dfrac{1}{z}$

13) $\dfrac{1}{x} - \dfrac{x+1}{xy} + \dfrac{x-2}{xz}$

14) $\dfrac{\dfrac{1}{y} - \dfrac{x}{z}}{\dfrac{1}{z} - \dfrac{1}{x}}$

15) $\dfrac{\dfrac{1}{st} - \dfrac{1}{w}}{\dfrac{1}{tw} - \dfrac{2}{s}}$

16) $\dfrac{y}{x} - \dfrac{x}{y}$

17) $\dfrac{4yz}{x^2} - \dfrac{2z}{xy^2} + \dfrac{1}{xyz}$

18) $\dfrac{\dfrac{1}{x} - \dfrac{x}{y}}{\dfrac{2y}{x} + \dfrac{2x}{y}} + \dfrac{x-y}{xyz}$

1.4 <u>Exponents</u>

Positive whole number exponents are a simple mathematical notation to represent repeated multiplication by the same factor. But you can also have negative numbers and 0 as exponents. Consider the table below.

3^4	$= 3\cdot3\cdot3\cdot3$	$=81$
3^3	$= 3\cdot3\cdot3$	$=27$
3^2	$= 3\cdot3$	$=9$
3^1	$= 3$	$=3$
3^0	$= 1$	$=1$
3^{-1}	$= \frac{1}{3}$	$= \frac{1}{3}$
3^{-2}	$= \frac{1}{3^2}$	$= \frac{1}{9}$
3^{-3}	$= \frac{1}{3^3}$	$= \frac{1}{27}$
3^{-4}	$= \frac{1}{3^4}$	$= \frac{1}{81}$

Exponents increase by 1.   Exponents decrease by 1.

Numbers are multiplied by 3.

Numbers are divided by 3.

See the pattern? Let's summarize: $3^n = \underbrace{3\cdot3\cdot3\cdot\;\cdots\;\cdot3\cdot3}_{n \text{ factors}}$, $3^{-n} = \dfrac{1}{3^n}$ and $3^0 = 1$. In general, given any number x and any positive integer n,

$$x^n = \underbrace{x\cdot x\cdot x\cdot\;\cdots\;\cdot x\cdot x}_{n \text{ factors}} \; ,$$

$$x^{-n} = \frac{1}{x^n}, \text{ for } x \neq 0,$$

and $$x^0 = 1, \text{ for } x \neq 0.$$

In all of these rules, the number x is called the <u>base</u>, while the number n is called the <u>exponent</u>. Notice that

$$x^m \cdot x^n \;=\; x^{m+n},$$

$$x^m/x^n \;=\; x^{m-n},$$

and

$$\left(x^m\right)^n = x^{m \cdot n}.$$

These rules are also true for values of m and n that are not integers. We will deal with such exponents later.

Example 1:   a) $\dfrac{4^2 - 1}{3^3 - 2^2} = \dfrac{16 - 1}{27 - 4} = \dfrac{15}{23}$

b) $5^{-1} + 3^{-1} = \dfrac{1}{5} + \dfrac{1}{3} = \dfrac{3 + 5}{15} = \dfrac{8}{15}$

c) $\dfrac{3 \cdot 8^2}{9 \cdot 8^3} = \dfrac{1 \cancel{3} \cdot 8^2}{3 \cancel{9} \cdot 8^3} = \dfrac{1}{3} \cdot \dfrac{8^2}{8^3} = \dfrac{1}{3} \cdot \dfrac{1}{8} = \dfrac{1}{24}$ ■

Example 2: $\dfrac{x^2 y^5}{x^{-3}} \div \dfrac{x^{-5} y^4}{x^3}$

$$= \left(\dfrac{x^2}{x^{-3}} y^5\right) \div \left(\dfrac{x^{-5}}{x^3} y^4\right)$$

$$= x^5 y^5 \div x^{-8} y^4$$

$$= \dfrac{x^5 y^5}{x^{-8} y^4} = x^{13} y \quad ■$$

Example 3: $\dfrac{1}{a^3} - \left(\dfrac{1}{a^5} - \dfrac{1}{a^2}\right)$

$$= \dfrac{1}{a^3} - \dfrac{1}{a^5} + \dfrac{1}{a^2}$$

$$= \dfrac{a^2 - 1 + a^3}{a^5} \quad ■$$

Example 4: $\left(\dfrac{x^{-2}}{x^8}\right)^{-2}$

$$= \left(x^{-2} x^{-8}\right)^{-2}$$

$$= \left(x^{-10}\right)^{-2} = x^{20} \quad ■$$

Exercises 1.4 Simplify the following expressions:

1) $\dfrac{4^{-1}5^2}{2^2 3^{-2}}$

2) $\dfrac{3^5 2^3}{4^2 3^3}$

3) $\dfrac{5^3}{3^{-1}5^2 + 4^{-1}5^3}$

4) $4^3 \left(\dfrac{1}{4}\right)^2 3^{-4}$

5) $\dfrac{1}{2^{-3}} - \dfrac{1}{2} + \dfrac{1}{5^{-2}}$

6) $x^2 y^{-2} z^3 x^{-2} y^3 z^5$

7) $\left(2^2\right)^{-1}$

8) $\left(x^2\right)^{34}$

9) $\left(\dfrac{1}{x^2}\right)^{34}$

10) Show by example that $\left(x^2 + y^2\right)^{30} \neq x^{60} + y^{60}$, i.e., find values for x and y so that the two sides are unequal for those values.

Simplify:

11) $\dfrac{x^{-1} y^2}{y^2 x^{-2}}$

12) $\dfrac{\left(x^2 y^{-3}\right)^2}{\left(y^{-3} x^{-2}\right)^{-2}}$

13) $\dfrac{x^2 y}{x^3} \div \dfrac{x^{-3} y^6}{y^4}$

14) $\dfrac{\dfrac{x^2 y^{-3}}{3z^2} - \dfrac{z^{-3} y^{-3}}{3x^2}}{\dfrac{x^{-4} y^2}{3z^{-2}}}$

1.5 <u>Roots</u> (also called radicals)

Definition: We say that the number x is a <u>square root</u> of the number y if $x^2 = y$. A <u>cube root</u> of the number y is a number x such that $x^3 = y$. (And x is a <u>fourth root</u> of the number y if $x^4 = y$, and so on.) If n is any positive integer, we say that the number x is an "nth root" of the number y if $x^n = y$. The number n is called the <u>order</u> of the root.

<u>Remarks for roots of even order:</u>

a) The number 16 has two square roots, 4 and -4. The "radical" symbol $\sqrt{}$ means the <u>positive</u> square root always! So, $\sqrt{16} = 4$, but $\sqrt{16} \neq -4$. We say that 16 has two square roots: $\sqrt{16}$ and $-\sqrt{16}$ (i.e., 4 and -4). The situation is similar for higher even-order roots. For example: $\sqrt[4]{16} = 2$, since $2^4 = 16$; also $\sqrt[6]{64} = 2$, $\sqrt[4]{81} = 3$, $\sqrt{\dfrac{1}{4}} = \dfrac{1}{2}$, and $-\sqrt{81} = -9$.

b) Since no real number multiplied by itself an even number of times can produce a negative number, negative numbers have no real roots of even order. So $\sqrt{-4}$ is not a <u>real</u> number. (Do you know what it is ?) Notice that we usually write $\sqrt{}$ instead of $\sqrt[2]{}$.

c) To sum up:

   i)   Every positive number x has two real nth roots if n is even – namely, $\sqrt[n]{x}$ and $-\sqrt[n]{x}$.

   ii)   Only positive numbers, and 0, have even-order roots that are real numbers.

<u>Remarks for roots of odd order:</u>

a) Things are a little different with odd-order roots. The number 8 has exactly 1 real cube root – namely, 2. So, $\sqrt[3]{8} = 2$. But notice $\sqrt[3]{-8} = -2$, because $(-2)^3 = -8$, and $\sqrt[3]{-27} = -3$, because $(-3)^3 = -27$. Get it? Multiplying a negative number by itself an odd number of times produces a negative number!

b) To sum up: all real numbers x have exactly one real nth root if n is odd – namely, $\sqrt[n]{x}$.

<u>An alternative notation</u> for $\sqrt[n]{a}$ is $a^{1/n}$. Both notations mean exactly the same thing. For example, $8^{1/3} = 2$, $25^{1/2} = 5$ (not -5), and $(-16)^{1/2}$ is not defined (as a real number).

Definition: We can also define <u>fractional exponents</u>: if $\dfrac{m}{n}$ is in lowest terms, and if n and a are such that $\sqrt[n]{a}$ makes sense, then we define $a^{m/n}$ to be $(\sqrt[n]{a})^m$, which equals $\sqrt[n]{a^m}$. If a is much larger than 1, it is usually easier to take the root first; it keeps the numbers down.

Example 1:   a) $8^{2/3} = (\sqrt[3]{8})^2 = 2^2 = 4$

b) $\left(\dfrac{-1}{27}\right)^{4/3} = \left(\sqrt[3]{\dfrac{-1}{27}}\right)^4 = \left(\dfrac{-1}{3}\right)^4 = \dfrac{1}{81}$

c) $(-32)^{4/5} = (\sqrt[5]{-32})^4 = (-2)^4 = 16$ ■

Laws of Exponents: Let r and s be any rational numbers. Let a and b be any real numbers. Then, in each of the following, if the expressions on both sides exist, they will be equal. (When might they not exist?)

1. $a^r \cdot a^s = a^{r+s}$

2. $\dfrac{a^r}{a^s} = a^{r-s}$

3. $(a^r)^s = a^{r \cdot s}$

4. $(ab)^r = a^r b^r$

5. $\left(\dfrac{a}{b}\right)^r = \dfrac{a^r}{b^r}$

6. $\left(\dfrac{a}{b}\right)^{-r} = \left(\dfrac{b}{a}\right)^r = \dfrac{b^r}{a^r}$

In Appendix A, we deal with real exponents that are not rational. The above laws of exponents are valid for irrational exponents also.

Exercises 1.5   Simplify the following as much as possible, using rational exponent notation where appropriate.

1) $\sqrt{144}$

2) $\sqrt[3]{-64}$

3) $\sqrt{\dfrac{1}{9}}$

4) $\sqrt[5]{-32}$

5) $\sqrt{\dfrac{4}{49}}$

6) $\sqrt[3]{\dfrac{8}{27}}$

7) $8^{5/3}$

8) $(-8)^{5/3}$

9) $(-32)^{2/5}$

10) $-(32)^{2/5}$

11) $\left(\dfrac{16}{9}\right)^{-3/2}$

12) $(.01)^{-3/2}$

13) $(.008)^{4/3}$

14) $2^{5/3} 2^{4/3}$

15) $8^{5/4} 4^{1/2}$

16) $\left(3^{2/3}\right)^{3/4}$

17) $\dfrac{2^{4/7}}{2^{1/2}}$

18) $3\left(27^{2/3}\right)^{1/2}$

19) $\dfrac{\left(2^{1/3}\right)^{2/5}}{\sqrt[5]{2}}$

20) $3^{1/3} 9^{1/3} + 2^{1/3} 16^{1/6}$

1.6 <u>Percent</u>

Percent, represented by the symbol %, means "per hundred." So 5% of 400 means 5 one-hundredths of 400, which is $\left(\dfrac{5}{100}\right)(400) = 20$. In general, x % of y is $\left(\dfrac{x}{100}\right)(y)$.

Example 1:   Find 15% of 90.

Solution: $\dfrac{15}{100} \cdot 90 = \dfrac{15 \cdot 9}{10} = \dfrac{135}{10} = 13.5$ ∎

Example 2:   Find 1% of 320.

Solution: $\dfrac{1}{100} \cdot 320 = \dfrac{320}{100} = 3.2$ ∎

Note that 1% of any number is always $\dfrac{1}{100} = .01$ of the number, so move the decimal point over two places to the left! (For example, 1% of 4567 is 45.67 and 1% of 378.2 is 3.782 .)

Example 3:   Find $\frac{1}{3}$% of 930.

Solution: Since 1% of 930 is 9.3, $\frac{1}{3}$% of 930 would be $\frac{1}{3}$ of 9.3 or 3.1.

Alternatively: $\dfrac{\frac{1}{3}}{100} \cdot 930 = \dfrac{930}{300} = \dfrac{93}{30} = \dfrac{31}{10} = 3.1$ ∎

Example 4:   Find 250% of 150.

Solution: $\dfrac{250}{100} \cdot 150 = \dfrac{250 \cdot 15}{10} = 25 \cdot 15 = 375$ ∎

Example 5:   A CD usually sells for $15.99. Shower Records is having a big sale offering 60% off. How many CD's can you get for $20, and exactly how much would you pay (assuming no sales tax)?

Solution: A 60%-off sale means that the CD's will cost 40% of their original price. Now 40% of $15.99 is

$$\frac{40}{100} \cdot (\$15.99) = \frac{4 \cdot (\$15.99)}{10} = \$6.396 = \$6.40 \quad \text{(rounded up to the}$$

nearest cent).

So you can buy three CD's at a cost of $19.20. ■

Example 6: The new federal budget calls for a 30% cut in the deficit to $147 billion. What would the deficit have been if it had not been cut?

Solution: Let x = the uncut deficit.
So 70% of x is 147 billion, that is

$$\frac{70}{100} \cdot x = 147 \text{ billion.}$$

Solving for x: $x = \dfrac{100}{70} \cdot (147 \text{ billion})$

$$= 210 \text{ billion.}$$

So the deficit would have been 210 billion dollars. ■

Exercises 1.6

1) Find 25% of 200.

2) Find 3.3% of 7.1.

3) The number 587 is 45% of what number?

4) The bookstore is having an inventory sale with a 25% reduction in prices. How much will a $16.99 sweatshirt cost you? How about a $1499.00 computer?

5) After a 40% cost reduction, a textbook that you purchased cost $20.40. What was the original price?

6) A hardware store is having a sale with a 20% reduction in prices. How much will a $239 table saw cost you?

7) After a 65% cost reduction, your new pet bird cost $84. What was the original price?

8) Joe heard that the government was selling foreclosed land at 35% off. He went to look at a farm going for the reduced price of $97,500. Joe bought the farm. How much did he save?

1.7 Scientific Notation

Decimal notation is OK for many numbers, but for really large or small numbers it is a big pain. Take the deficit (please!): 210 billion = 210,000,000,000 (in decimal form). But notice that all the zeros really represent multiplication by 10 so that

$$210,000,000,000 \ = \ 21 \times 10^{10} \ = \ 2.1 \times 10^{11}.$$

This last expression using the factors of 10 is called scientific notation. Notice that scientific notation calls for:

$$\pm \begin{bmatrix} \text{a number greater than or equal} \\ \text{to 1, but less than 10, written} \\ \text{in decimal form} \end{bmatrix} \times 10^{\left(\begin{array}{c} \text{some integer, positive} \\ \text{or negative, or 0} \end{array} \right)}$$

For example, you would write .0000536 in scientific notation as 5.36×10^{-5}. You would write 6437.8 as 6.4378×10^{3}. Just count the number of positions you have to move the decimal point so that the resulting number is between 1 and 10. If you have to move the decimal point left n times, the exponent is n; if you move it n places to the right, the exponent is $-n$.

Remark: On most calculators, very large or small numbers are given in something like scientific notation. For example, the number 3.28×10^{-9} may look like 3.2800000 E -09, or 3.28 $^{-09}$.

Example 1:   The number of molecules in a mole (a mole??) of gas is called Avogadro's number. It is equal to 6.023×10^{23} . Write it in decimal [non-scientific] form.

Solution: $6.023 \times 10^{23} \ = \ 602,300,000,000,000,000,000,000$
See why scientific notation is so handy! ■

Example 2:   The speed of light (in a vacuum) is 186,000 miles per second. How big is a light year (the distance light travels in 1 year)? Express the answer in scientific notation, rounded to three digits.

Solution: First, the speed of light written in scientific notation is 1.86×10^{5}. Now there are 365 days a year, so there are
365 days × 24 (hr/day) = 8760 hours per year, and
365 (days/year) × 24 (hr/day) × 3600 (sec/hr) = 31536000 seconds per year.
In scientific notation, this number is 3.1536×10^{7} seconds.

We want to compute distance = velocity × time, so

total distance = $(1.86 \times 10^5) \times (3.1536 \times 10^7)$.

This raises the issue of how to multiply two numbers written in scientific notation. It's really quite easy.

$$(x_1 \times 10^{e_1}) \times (x_2 \times 10^{e_2}) = x_1 \times x_2 \times 10^{e_1} \times 10^{e_2}$$
$$= (x_1 \times x_2) \times 10^{e_1 + e_2}$$

However, we must make sure that $|x_1 \times x_2|$ is between 1 and 10. If it is not, we have to move the decimal point and adjust the exponent. So

$$(1.86 \times 10^5) \times (3.1536 \times 10^7) = (1.86 \times 3.1536) \times 10^{12}$$
$$= 5.865696 \times 10^{12}$$
$$= 5.87 \times 10^{12} \text{ miles (rounded to 3 digits)} \quad \blacksquare$$

Remark: In this case the answer is in scientific form. However, consider the product

$$(3.0 \times 10^3) \times (4.0 \times 10^7) = 12.0 \times 10^{10}$$
$$= 1.2 \times 10^{11}.$$

Here we have the extra step of moving the decimal point over.

Exercises 1.7

1) Express the following numbers in scientific notation, rounded to three digits.

   a) 382935.9938   b) -0.000724 c) 3.000001   d) 200.001

2) Compute the following and express in scientific notation, rounded to three digits. You may use your calculator.

   a) $(2.35 \times 10^5) \times (4.032 \times 10^2)$ b) $(-6.15 \times 10^{-2}) \times (5.032 \times 10^6)$

   c) $(-5.001 \times 10^{-2}) \times (-7.001 \times 10^{-99})$ d) $\dfrac{3.24 \times 10^2}{4.23 \times 10^3}$

   e) $\dfrac{-1.33 \times 10^{-2}}{7.9 \times 10^5}$ f) $(3.82 \times 10^{-1})^3$

Chapter 2

Completing the Square

A quadratic expression in x is $ax^2 + bx + c$. It's a polynomial of degree 2. One important way of changing the form of a quadratic is called <u>completing the square</u>. It is one of the most frequently used methods of changing the form of an expression to suit a particular purpose. It is used for graphing circles, ellipses, parabolas, and hyperbolas, for deriving the quadratic formula and integrating certain functions, and for many other purposes. You'll meet them soon enough. We'll do an example first, and then look at the general method.

Example 1:   Complete the square for $f(x) = x^2 + 8x + 12$.

Solution: (First, notice that the x^2 coefficient is 1. If it is not, it must be factored out.) We take 8, the coefficient of x, take half of the 8 to get 4, square the 4 to get 16, which we add (<u>and subtract</u>), to get

$$f(x) = (x^2 + 8x + 16) + (12 - 16).$$

The first term is a perfect square. (That was the whole idea – and that's why it's called "completing the square.") So

$$f(x) = (x+4)^2 - 4. \blacksquare$$

Here's the general method for completing the square of $f(x) = ax^2 + bx + c$:

a)   If $a \neq 1$, factor out the a from the first two terms to get $f(x) = a\left(x^2 + \dfrac{b}{a}x\right) + c$.

b)   Take half of the coefficient of the resulting x term, and square it.

c)   Add and subtract that number (inside the parentheses if $a \neq 1$).

d)   Rewrite $f(x)$ as the sum of a perfect square and a number.

Example 2:   Complete the square for $f(x) = x^2 - 3x + 4$.

Solution: Here, the x^2 coefficient is 1, which means we can skip the first step. The x coefficient is -3, half of it is $-\dfrac{3}{2}$ and squaring it gives $\dfrac{9}{4}$. Adding and subtracting $\dfrac{9}{4}$ gives us

$$f(x) = \left(x^2 - 3x + \frac{9}{4} \right) + \left(4 - \frac{9}{4} \right).$$

The first term is the perfect square of $x - \dfrac{3}{2}$, so

$$f(x) = \left(x - \frac{3}{2} \right)^2 + \frac{7}{4}. \quad \blacksquare$$

Example 3:   Complete the square for $f(x) = 4x^2 + 20x - 100$.

Solution: Now $a = 4$, so we must first factor it out:

$$f(x) = 4(x^2 + 5x) - 100.$$

The coefficient of the x term is 5, halving and squaring gives $\dfrac{25}{4}$ as the term to add and subtract, so

$$f(x) = 4\left(x^2 + 5x + \frac{25}{4} - \frac{25}{4} \right) - 100.$$

(Notice the $\dfrac{25}{4}$ is subtracted <u>inside</u> the bracket.)

$$= 4\left(x^2 + 5x + \frac{25}{4} \right) - 25 - 100$$

$$= 4\left(x + \frac{5}{2} \right)^2 - 125 \quad \blacksquare$$

Sometimes we need to complete the square in an equation. We may also need to complete the square in more than just one variable. Check the next example.

Example 4:   Complete the square in x and y for $x^2 - 4x + y^2 + 6y = 2$.

Solution: Since half of the x-coefficient is –2, when squared equals 4, we must first add 4 to both sides to complete the square in x. Next, half of the y-coefficient is 3, when squared equals 9, so we add 9 to both sides giving

$$(x^2 - 4x + 4) + (y^2 + 6y + 9) = 2 + 4 + 9.$$

(Notice that instead of adding and subtracting on the left side, we added the same amount to both sides, which amounts to the same thing.)

So $(x-2)^2 + (y+3)^2 = 15$.

[By the way, this equation is that of a circle of radius $\sqrt{15}$ centered at the point $(2,–3)$.] ∎

The particular problem you solve will determine if you choose to add the number to both sides of the equation, or just add and subtract the number on one side. It all amounts to the same result.

Example 5:   Complete the square in x and y for $4x^2 - 9y^2 + 8x + 18y - 25 = 0$.

Solution: First regroup terms and factor out the coefficients of the quadratic terms:

$$4(x^2 + 2x) - 9(y^2 - 2y) - 25 = 0$$

Now add and subtract appropriate constants: in this case both are 1.

$$4(x^2 + 2x + 1 - 1) - 9(y^2 - 2y + 1 - 1) - 25 = 0$$

Upon simplifying:

$$4(x + 1)^2 - 4 - 9(y - 1)^2 + 9 - 25 = 0$$

or $4(x + 1)^2 - 9(y - 1)^2 = 20$. ∎

Example 6:   Complete the square in x and y for $x^2 - \pi x + 2y^2 - y = 0$.

Solution: For the quadratic in x, we need to add and subtract $\dfrac{\pi^2}{4}$, while for the y terms we need to first factor out the 2 and then add and subtract $\dfrac{1}{16}$ giving

$$x^2 - \pi x + \frac{\pi^2}{4} - \frac{\pi^2}{4} + 2(y^2 - \frac{1}{2}y + \frac{1}{16} - \frac{1}{16}) = 0.$$

Simplifying gives

$$\left(x - \frac{\pi}{2}\right)^2 + 2\left(y - \frac{1}{4}\right)^2 = \frac{1}{8} + \frac{\pi^2}{4} = \frac{1 + 2\pi^2}{8}. \quad \blacksquare$$

Example 7:   Complete the square to solve $ax^2 + bx + c = 0$.

Solution: $0 = ax^2 + bx + c = a\left(x^2 + \frac{b}{a}x\right) + c$

$$= a\left(x^2 + \frac{b}{a}x + \frac{b^2}{4a^2} - \frac{b^2}{4a^2}\right) + c$$

$$= a\left(x^2 + \frac{b}{a}x + \frac{b^2}{4a^2}\right) - a\frac{b^2}{4a^2} + c$$

$$= a\left(x + \frac{b}{2a}\right)^2 - \frac{b^2}{4a} + c$$

$$= a\left(x + \frac{b}{2a}\right)^2 - \frac{b^2 - 4ac}{4a}$$

So, $a\left(x + \frac{b}{2a}\right)^2 = \frac{b^2 - 4ac}{4a}.$

Let's "peel the onion" to solve for x. First get rid of a:

$$\left(x + \frac{b}{2a}\right)^2 = \frac{b^2 - 4ac}{4a^2}$$

Take square roots:

$$x + \frac{b}{2a} = \pm\sqrt{\frac{b^2 - 4ac}{4a^2}} = \pm\frac{\sqrt{b^2 - 4ac}}{2a},$$

and so

$$x = -\frac{b}{2a} \pm \frac{\sqrt{b^2 - 4ac}}{2a}$$

$$= \boxed{\frac{-b \pm \sqrt{b^2 - 4ac}}{2a}}.$$

Look familiar? ■

Exercises 2.1

1) Complete the square for the following expressions:

a) $f(x) = x^2 - 6x + 15$ b) $h(y) = y^2 + 5y$

c) $g(s) = s^2 + 2s - 8$ d) $k(x) = 2x^2 - 2x + 5$

e) $f(x) = 3x^2 - 7x + 1$ f) $w(x) = \pi x^2 + 2x$

2) Complete the square for the following equations:

a) $x^2 - 3x - 17 = 0$ b) $-3x^2 - 6x + 15 = 0$

3) Complete the square in both x and y for the following equations:

a) $x^2 + 3x + 2y^2 - 8y = 0$ b) $3x^2 + 6x - 2y^2 - 8y = -11$

c) $-x^2 + 4x + y^2 - 16y = 40$ d) $-9x^2 + 36x - 4y^2 - 8y = 0$

e) $x^2 + y^2 - 6x + 10y + 34 = 0$

The graph of this last example is called a degenerate circle. (Can you figure out why?)

Chapter 3

Power Functions, Shifting and Scaling

3.1 Functions

Before delving into power functions and their graphs, let's first talk a little about functions in general. Functions, as you are learning, are rules or operations that take one number, say x, and associate with it another number, call it y. Some functions are given as tables, others as graphs, and others as analytical expressions – for example, $f(x) = 2x + \sqrt{3x - 1}$. In the latter case, for a particular value of x we substitute that value into the formula for the function and out comes the appropriate y value. The set of all x's is called the domain, and the set of all the y's, or $f(x)$ values, is called the range.

Example 1: Suppose $f(x) = 2x^2$; what is $f(4)$, $f(-3)$, $f(4+h)$, $f(x+\Delta x)$, and $f\left(\sqrt{\dfrac{x}{2}}\right)$? (Think of Δx as a variable that happens to looks a little different. Get used to it !)

Solution: Simply replace x in the expression by the appropriate number. So:

$$f(4) = 2 \cdot 4^2 = 2 \cdot 16 = 32,$$

$$f(-3) = 2 \cdot (-3)^2 = 2 \cdot 9 = 18,$$

$$f(4+h) = 2(4+h)^2 = 2(16 + 8h + h^2) = 32 + 16h + 2h^2,$$

$$f(x+\Delta x) = 2(x+\Delta x)^2 = 2(x^2 + 2x\Delta x + (\Delta x)^2)$$
$$= 2x^2 + 4x\Delta x + 2(\Delta x)^2$$

and $\quad f\left(\sqrt{\dfrac{x}{2}}\right) = 2\left(\sqrt{\dfrac{x}{2}}\right)^2 = 2 \cdot \dfrac{x}{2} = x .$ ∎

As we said above, functions can be given in several ways. Here are some of the functions we'll meet:

a) The absolute value function is defined as

$$|x| = \begin{cases} x & \text{if } x \geq 0 \\ -x & \text{if } x < 0 . \end{cases}$$

Remark: Even though "$-x$" may LOOK negative, it is <u>NOT</u> negative if $x < 0$. For example, if $x = -5$, then $-x = -(-5) = 5$. Also notice that $\sqrt{x^2} = |x|$. (Verify this for $x = 3$, and $x = -3$.)

b) The sine function, denoted $\sin x$, is given by the graph below.

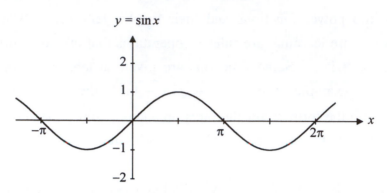

The graph continues on indefinitely on both sides in a similar fashion. It is called <u>periodic</u> with period 2π since the entire behavior of the function is given on any interval of length 2π – e.g., the interval $[0, 2\pi]$. See Chapter 6 for more on this function and other so-called trig functions, like the next example.

c) The cosine function, denoted $\cos x$, is also given by a graph.

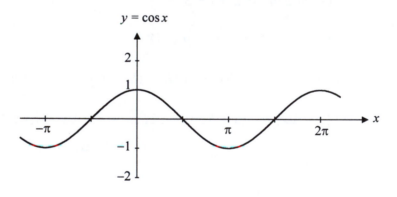

If you have not had trig functions before, don't panic. Just think of them as being <u>defined by these two graphs</u>.

Note: a) $f(x) > 0$ wherever the graph of $f(x)$ is above the x-axis.

b) $f(x) = 0$ wherever the graph of $f(x)$ crosses or touches the x-axis.

c) $f(x) < 0$ wherever the graph of $f(x)$ is below the x-axis.

Example 2:   Solve $\sin x < 0$ for x in $[-\pi, \pi]$.

Solution: From the graph we see that the function $\sin x$ is strictly below the x-axis on the open interval $(-\pi, 0)$, and on or above the x-axis at all other points. Hence the set of solutions is $(-\pi, 0)$. ∎

Exercises 3.1

1) Let $f(x) = x^3 + 2x$. Determine:
    a) $f(2)$ b) $f(3)$ c) $f(x+h)$
    d) $f(2x)$ e) $f(-x)$ f) $f(2+\Delta x)$

2) Evaluate $|x|$ at $x = 0, \pm 1, \pm 2$, and sketch the graph.

3) Evaluate
    a) $|3 - 4|$ b) $|4 - 3|$ c) $\sin \pi$

    d) $\cos 0$ e) $\left(\sin \dfrac{\pi}{2}\right)(\cos \pi)$

4) Suppose $f(x) = x^4 - \cos x$, then determine:
    a) $f(0)$ b) $f(\pi)$ c) $f(x+h)$

    d) $f\left(\dfrac{\pi}{2}\right)$ e) $f(-x)$

5) From the graph above, solve the equation $\sin x = 0$. (As usual, this means: find all values of x for which the equation holds.)

6) Solve $\cos x \geq 0$ for $0 \leq x \leq 2\pi$. (Again: find all x that make the inequality true.)

7) Solve: $|x| - 2 < 0$.

8) Solve $x^2 - 9 > 0$.

3.2 Linear Functions

What do the functions x, $2x$, or in general, mx (for m being some constant) look like? We can use plotting of points to get the following:

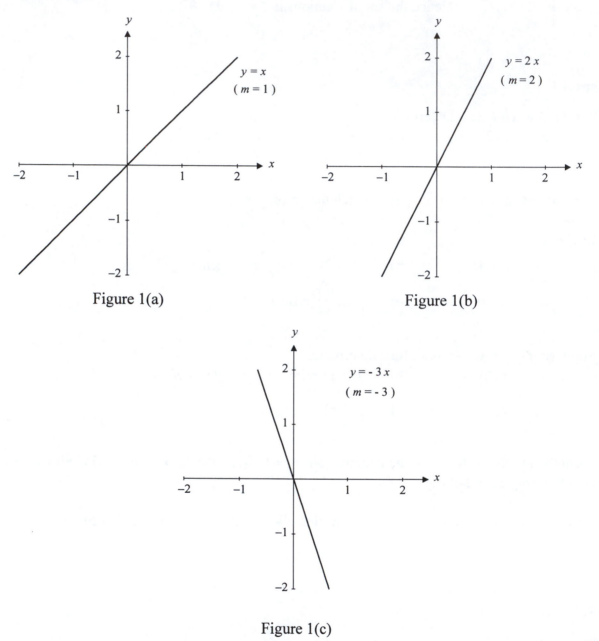

Figure 1(a)   Figure 1(b)

Figure 1(c)

These figures show the graph of the function $y = mx$ for $m = 1, 2$, and -3. This "family" of functions represents all straight lines passing through the origin, with one exception: the vertical line. The number m, as you may know, is called the <u>slope</u> of the line. As we go from left to right (our usual assumption), if m is positive, the line rises; if it is negative, the line sinks. The larger the magnitude of the slope, $|m|$, the steeper the line becomes. (What happens when the slope is a really big positive number?) Figure 2 illustrates several of these graphs on the same axes for comparison purposes.

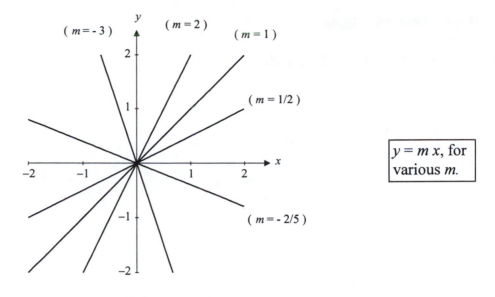

Figure 2

<u>Remark:</u>   It helps, whenever possible, if you use the same scale on the x and y axes, meaning the distance from 0 to 1 is the same on the two axes. Sometimes that is really not feasible, as in exercise 5 below. In any event, always mark your scales – i.e., show where 1, 2, etc. (or other appropriate values) are.

The general equation of a line is $y = mx + b$. We will see in Section 3.4 that the additive term b amounts to a vertical shifting up or down. Notice that if we let $x = 0$ (which corresponds to the y-axis), then $y = b$. In other words, the line intersects the y-axis at $y = b$. For this reason, the number b is called the y intercept.

Exercises 3.2 Graph the following functions, by plotting two points each. (Why is it enough to plot only two points ?)

1) $f(x) = \dfrac{1}{3} x$

2) $y = -1.5\, x$

3) $g(x) = .6\, x$

4) $f(x) = 5\, x$

5) $y = 1000\, x$

3.3 <u>Other Power Functions</u>

Part A: $y = x^r$, with $r = 2, 4, 6$, etc.

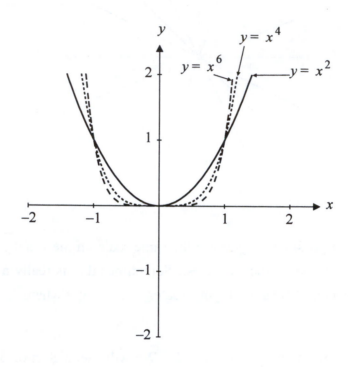

Figure 3

Notice how the points $x = 1$ and $x = -1$ are "pivotal"? Consider $x > 0$, or the right side of the graph. For $x < 1$, the relative position of the curves is opposite to what it is when $x > 1$. Well, let's think about it: for $x < 1$, the more we multiply it by itself the smaller it gets. The higher the power, the smaller the number gets. When $x > 1$, the opposite is true. The higher the power, the more we multiply by a number which is greater than 1, hence the larger the number gets. Notice that these graphs seem symmetric about the y-axis. In fact, they are. It doesn't matter whether we insert x or $-x$ into an even power function; the answer is the same. (Can you show this?) This shows <u>symmetry about the y-axis</u>.

Part B: $y = x^r$, with $r = 3, 5, 7$, etc.

Symmetry through the origin is exhibited by the odd power functions.   Figure 4 shows three members of this little family.

As in Figure 3, the points $x = 1$ and $x = -1$ are pivotal in that the relative positioning of the graphs changes there.   Now, however, the graphs are symmetric through the origin.   That is, if the point (x , y) is on the graph, then so is $(-x , -y)$.

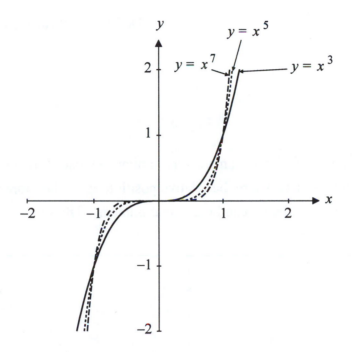

Figure 4

Part C : $y = x^r$, with $r = \frac{1}{2}, \frac{1}{4}, \frac{1}{6}$, etc., and $r = \frac{1}{3}, \frac{1}{5}, \frac{1}{7}$, etc.

What if r is a positive fraction of the form $\dfrac{1}{n}$ where n is a whole number 1, 2, 3, etc.?   What does $y = x^{1/2}$, $y = x^{1/3}$, etc. look like?   Well, as it turns out, this family of graphs also splits up into two groups – namely, the even n graphs and odd n graphs.   Within each group, the graphs display similar behavior.   For $y = x^{1/2}$ and $x^{1/4}$, the graphs look like that in Figure 5.   Since $y = x^{1/n}$ is equivalent to the nth root, if n is even x must not be negative.

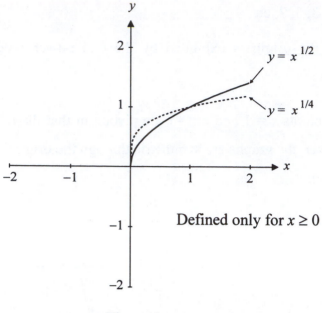

Figure 5

The other graphs of $y = x^{1/n}$ for n even are very similar to those shown in Figure 5, and as is the case in Figures 3 and 4, the point $x = 1$ is where the relative positioning of the graphs changes. If n is odd – say for example $y = x^{1/3}$, or $y = x^{1/5}$ – then x can be any real number. The graphs of these functions are shown in Figure 6.

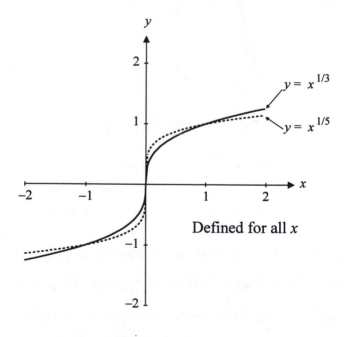

Figure 6

The other graphs of $y = x^{1/n}$ for n odd are very similar to those shown in Figure 6, with the same ordering situation.

To sum up Parts A, B, C: for $x \geq 0$, $y = x^r$ is shown in Figure 7.

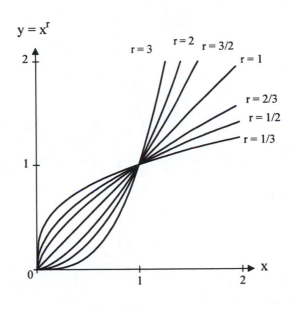

Figure 7

All of these are examples of <u>power functions</u>. Notice that $x^{2/3}$ is calculated by squaring $x^{1/3}$.

Part D: $y = x^r$, with $r = -1, -3$, etc. , and $-2, -4$, etc.

What if r is a negative integer? Can you guess that the situation will divide up into two cases, r even or odd? Here's the situation for r odd: $\dfrac{1}{x}, \dfrac{1}{x^3}$, etc. (Recall that x^{-3} means $\dfrac{1}{x^3}$, etc.)

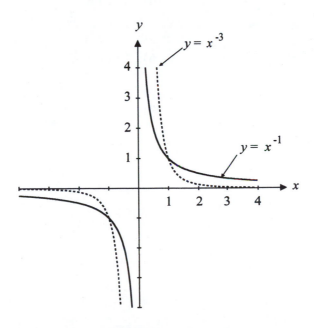

Figure 8

For $r = -2$ and -4, the graph is shown in Figure 9.

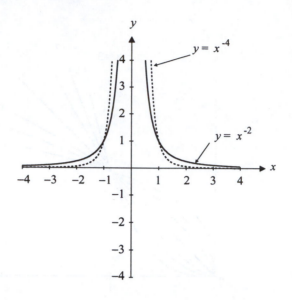

Figure 9

The functions x^{-3}, x^{-5}, etc. all basically look like x^{-1}, and x^{-4}, x^{-6}, etc. basically look like x^{-2}. The points $x = 1$ and $x = -1$ are again important since the relative positioning changes there.

Exercises 3.3 Using graph paper, calculate and plot exactly at least four points for each of the following:

1) $f(x) = \sqrt{x}$

2) $f(x) = x^3$

3) $g(x) = \sqrt[3]{x}$

4) $f(x) = \dfrac{1}{x}$

5) $g(x) = \dfrac{1}{x^2}$

6) $w(x) = x^{-3}$

7) $f(x) = x^{2/3}$

8) $f(x) = x^{3/2}$

3.4 Shifting Up or Down

How does the graph of $y = x^2 + 2$ compare to the graph of $y = x^2$? Well, all the y-coordinates of the first graph are 2 bigger than those of the second graph. But y is the altitude, or height, of the point (x,y). So to go from the graph of $y = x^2$ to $y = x^2 + 2$, just push up the graph a distance of 2, as shown in Figure 10.

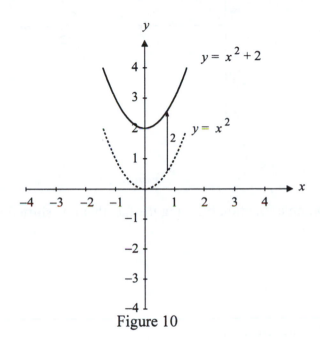

Figure 10

Also consider these examples:

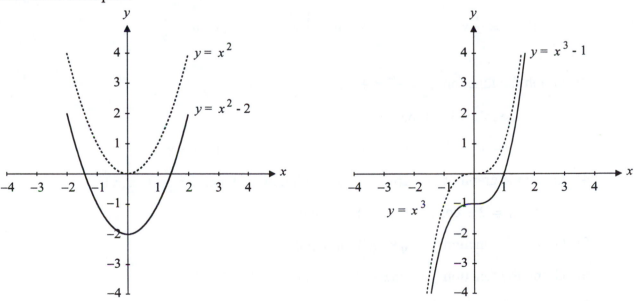

Figure 11(a)   Figure 11(b)

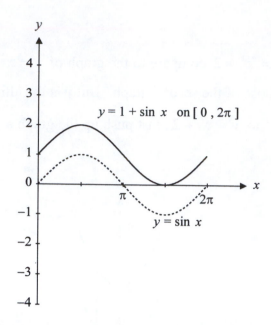

Figure 12

In each case, we are graphing a function by using the fact that it is shifted up or down from the graph of a function that we know.

Exercises 3.4

1) Graph the following functions:

   a) $y = \dfrac{1}{2}x$

   b) $y = \dfrac{1}{2}x + 2$

   c) $y = \dfrac{1}{2}x - 1$

2) Graph the function $y = -2x + 3$.

3) Graph the following functions:

   a) $y = x^2 - 4$

   b) $y = x^2 + \pi$

   c) $y = x^{-2} + 1$

4) Given the graph of $y = \cos x$, graph the following functions on $[-\pi, \pi]$:

   a) $y = 2 + \cos x$

   b) $y = \cos x - 4$

5) Graph the function $y = \sqrt{x} + 2$ for $x \geq 0$.

6) Graph the function $y = \sin x - 2$ for x in $[0, 2\pi]$.

7) Graph $|x|$ and $|x| - 2$.

3.5 Shifting Left or Right

Consider the following three graphs:

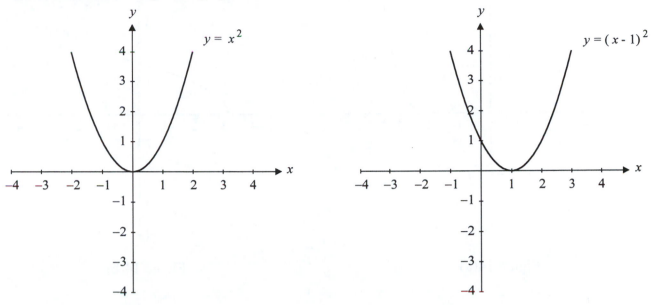

Figure 13(a)   Figure 13(b)

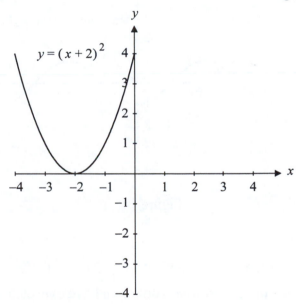

Figure 13(c)

(Don't take <u>our</u> word for it; test these by plotting points for $x = 0, \pm 1, \pm 2, \pm 3$.)

Notice that going from $f(x) = x^2$ to $f(x-1) = (x-1)^2$ we needed to shift the graph to the <u>right</u> a distance of 1, and going from $f(x) = x^2$ to $f(x+2) = (x+2)^2$ shifted the graph to the <u>left</u> a distance of 2. <u>This is true for all functions $f(x)$</u>, not just $f(x) = x^2$. In general, if we suppose a is some positive number, then the graph of $f(x - a)$ is the graph of $f(x)$ moved to the right a distance of a, and the graph of $f(x + a)$ is the graph of $f(x)$ moved to the left a distance of a. For instance, check the following graphs:

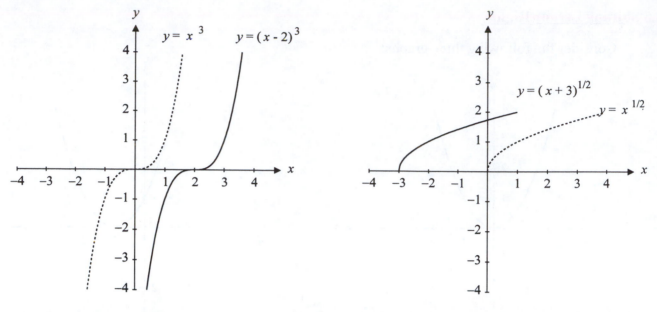

Figure 14(a)   Figure 14(b)

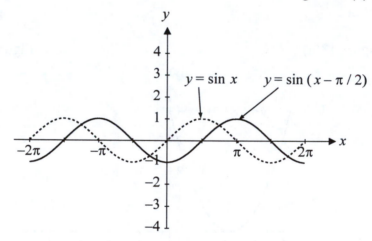

Figure 14(c)

<u>Remarks:</u>   a)   Notice in Figure 13 that the "vertex" of $(x + 2)^2$ is at -2, while the vertex of $(x - 1)^2$ is at $+1$, which may be opposite to what you might have expected, but plotting a few points proves it.

b)   It is important to see that, for example, $(x + 3)^2$ is very different from $x^2 + 3$. If we start from the graph of x^2, $(x + 3)^2$ moves it <u>3 units to the left</u>, while $x^2 + 3$ moves it <u>up a distance of 3 units</u>.

Exercises 3.5

1) Graph x^2, $(x + 1)^2$, $(x - 1)^2$, by plotting five points for each.

2) Graph the following functions:

    a) $f(x) = (x - 3)^2$ b) $y = (x + \pi)^2$ c) $y = \dfrac{1}{x - 1}$

3) Graph the following functions:

    a) $y = (x + 3)^3$ b) $y = \sqrt[3]{x + 1}$ c) $y = -\dfrac{1}{(x - a)^4}$, where $a > 0$.

4) Graph the function $y = \sqrt{x - 4}$ for $x \geq 4$.

5) Graph the function $y = \cos(x - \pi)$ for $x \in [-2\pi, 2\pi]$.

3.6 <u>Vertical Scaling</u> (Stretching and compressing)

Consider the graph of $y = \sin x$ on the interval $[0, 2\pi]$ shown in Figure 15. How do you stretch it away from the x-axis so it is twice as high above the axis, and twice as low below the axis? Simple: y needs to be twice as big, so $y = 2\sin x$ will do it.

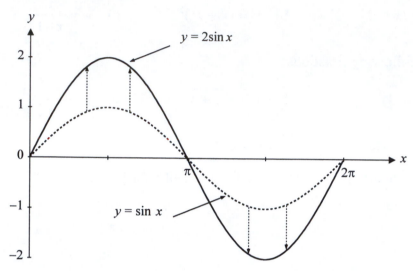

Figure 15

To compress the graph so that it lies between $y = -1/2$ and $y = 1/2$, simply multiply the function $\sin x$ by 1/2. Figure 16 illustrates $y = \dfrac{1}{2}\sin x$.

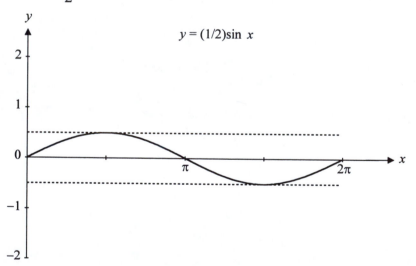

Figure 16

In general, when given the graph of $f(x)$, the graph of $k \cdot f(x)$ is that of $f(x)$ stretched if $k > 1$ and compressed if $0 < k < 1$. What happens if $k < 0$? Well, let's see. The graph of $y = -\sin(x)$ is shown in Figure 17. It's the mirror reflection of the graph of $y = \sin x$ about the x-axis.

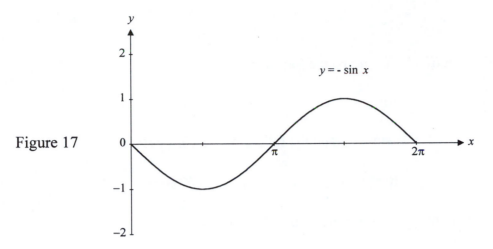

Figure 17

In general when given the graph of $f(x)$, the graph of $k \cdot f(x)$ for $k < 0$ is that of $f(x)$ flipped and stretched if $k < -1$, and flipped and compressed if $-1 < k < 0$, and just flipped if $k = -1$. Consider the following examples:

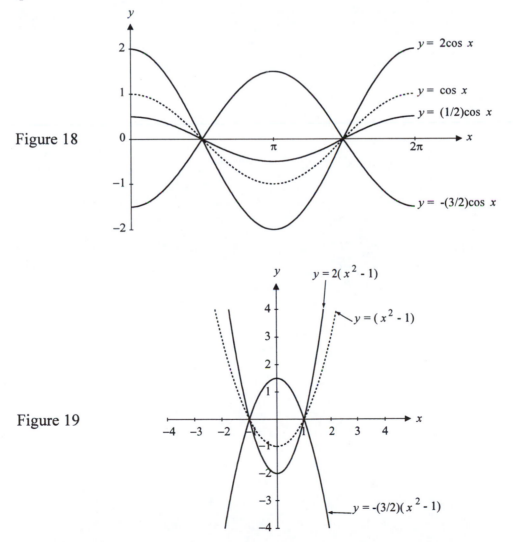

Figure 18

Figure 19

Exercises 3.6

1) From the graph of $y = x^2$, graph the following:

    a) $y = 2x^2$ b) $y = -x^2$ c) $y = \dfrac{1}{3}x^2$ d) $y = -\dfrac{1}{2}x^2$

2) From the graph of $y = \sin x$, graph the following on $[0, 2\pi]$:

    a) $y = 3\sin x$ b) $y = -\dfrac{1}{2}\sin x$ c) $y = -2\sin x$

3) Graph the function $y = -\sqrt{x}$ for $x \geq 0$.

3.7 Horizontal Scaling

When graphing sin x, as x goes from 0 to 2π, sin x goes from 0 to 1, to -1, back up to 0:

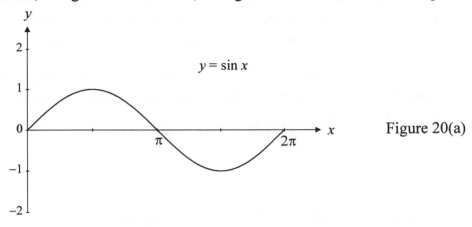

Figure 20(a)

(Sin x is said to go through <u>one cycle</u> on $[0, 2\pi]$, because the rest is repetition.) Old news! Ho hum! Now, what happens to sin $2x$? Well, as x goes from 0 to 2π, $2x$ goes from 0 to 4π, and sine of it (i.e., sin $2x$) will have the time to go through two cycles:

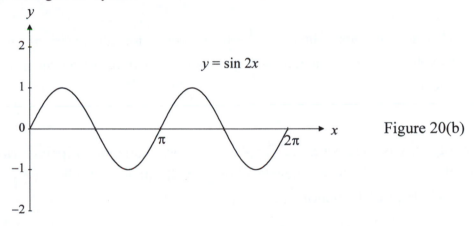

Figure 20(b)

On the other hand, for $\sin\frac{1}{2}x$, as x goes from 0 to 2π, $\frac{1}{2}x$ goes from 0 to π (i.e., through only half a cycle):

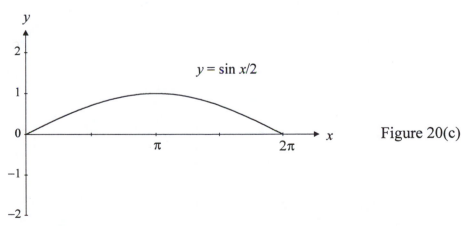

Figure 20(c)

In general here's how you go from $y = \sin x$ to $y = \sin a x$. (In fact, the same is true for any function $f(x)$.)

(a)   If $a > 1$, you compress the graph toward the y-axis. If a is a positive integer, then the graph of $y = \sin a x$ has a complete <u>oscillations</u>, or cycles, in the interval $[0,2\pi]$.

(b)   If $0 < a < 1$, you stretch the graph away from the y-axis. The graph of $y = \sin a x$ has a complete oscillations, or cycles, in the interval $[0,2\pi]$. For example, if $a = \frac{1}{5}$, we would have one-fifth of the total cycle.

Don't forget:

(a)   You can't "factor out" the 2 and say that $\sin(2x)$ and $2\sin(x)$ are the same thing even if the devil tempts you to. (How simple life would be!) Test it with $x = \dfrac{\pi}{2}$.

(b)   You also can't say that $\sin(x+y)$ and $\sin x + \sin y$ are equal, except by lying or fooling yourself.

<u>Remark:</u>   This scaling, horizontal and vertical, works for all functions, not just trig functions. However, the cyclic nature of trigonometric functions makes them perfect to illustrate these effects.

By the way, just as there is vertical flipping, so is there horizontal flipping. How do you flip the graph of $f(x)$ about the y axis to get a new function we will call $g(x)$? Well, at $x = a$, you set $g(a)$ equal to $f(-a)$, as shown in Figure 21. Hence $g(x) = f(-x)$.

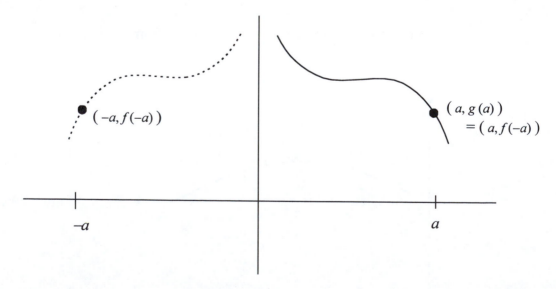

Figure 21

Example 1:

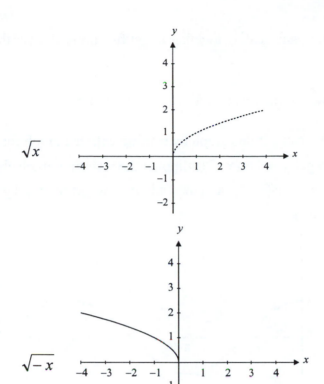

Exercises 3.7

1) Graph the following functions over the interval $[0, 2\pi]$:

   a) $y = \sin 4x$ b) $y = \cos 3x$ c) $y = \sin \frac{1}{4} x$

2) Graph $y = \sqrt{x}$ and $y = \sqrt{4x}$ on the same graph.

3) Graph $y = \dfrac{1}{x}$ and $y = \dfrac{1}{2x}$ on the same graph.

4) Graph $\sin(-x)$ on $[0, 2\pi]$, by first graphing $\sin x$ on $[-2\pi, 0]$

3.8 Combinations

We can put all this shifting and scaling together to get functions all over the place.

Example 1:   Sketch the graph of $y = 1 - \cos(x/2)$ on $[0, 2\pi]$.

Solution: We can get this graph by starting with the graph for $y = \cos x$, stretching this to get $y = \cos(x/2)$, flipping the resulting graph about the x-axis to get $y = -\cos(x/2)$, and then adding 1 to get the final graph $y = 1 - \cos(x/2)$. This gives;

a)
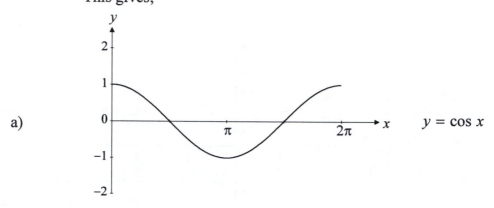
$y = \cos x$

b)
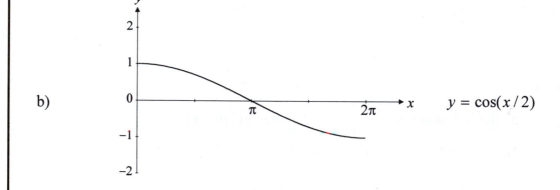
$y = \cos(x/2)$

c)
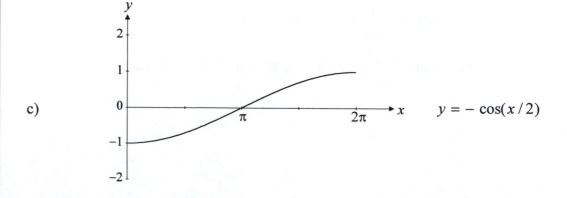
$y = -\cos(x/2)$

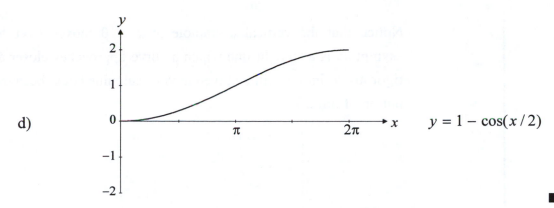

d) $y = 1 - \cos(x/2)$

Example 2:   Graph the function $y = \dfrac{1}{(x-1)} + 2$.

We graph this by starting with the graph for $y = \dfrac{1}{x}$, shifting this right one

unit to get $y = \dfrac{1}{(x-1)}$, and then adding 2 to get the final graph

$y = \dfrac{1}{(x-1)} + 2$.

Solution:

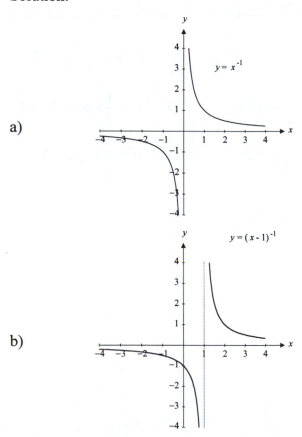

a)

b)

Notice that the vertical asymptote at $x = 0$ moves over to $x = 1$. (An asymptote is a straight line which a curve approaches closer and closer. The rigorous definition will be given in your calculus book, because it involves the notion of limit.)

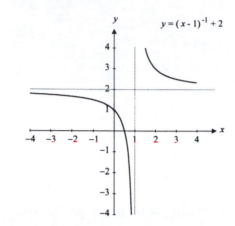

$$y = (x-1)^{-1} + 2$$

c)

Notice that the horizontal asymptote at $y = 0$ moves up to $y = 2$. ∎

Exercises 3.8 Graph the following functions:

1) a) $y = x^2$ b) $y = x^2 - 1$ c) $y = \frac{1}{2}(x^2 - 1)$ d) $y = \frac{1}{2}(x^2 - 1) + 2$

2) a) $y = \frac{1}{x}$ b) $y = \frac{1}{x-2}$ c) $y = \frac{3}{x-2}$ d) $y = \frac{3}{x-2} - 1$

3) $y = -2\sin(2x)$ 4) $y = -\frac{1}{2}\cos(2x)$ 5) $y = \frac{1}{2}\sin(x/2) + 1$

6) $y = (x-1)^2 - 1$ 7) $y = \frac{1}{x+1} + 2$ 8) $y = 2 - \sqrt{x-2}$

9) $y = (x-1)^3 + 2$ 10) $y = 1 - \frac{1}{(x-\pi)^4}$

Chapter 4

Factoring, Rationalizing, Extracting

In most limit problems that you'll meet, there will be a fractional expression in which the numerator and denominator have a common factor that needs to be canceled out. Moral: you need to know how to factor to be a player in this game. It isn't always obvious how you should factor, but in your factoring toolbox you've got four basic methods that should be tried in this order:

1) Common factors

2) Special formulas

3) Grouping

4) The factor theorem

Recall that factoring an expression means writing it as a product.

4.1 Common Factors

The easiest tool to use is to take out all factors that are common to all the terms of the expression.

Example 1: Factor $3x^2y^3 + 15xy^4 - 21x^3y^2$.

Solution: Notice that all three terms share a factor of 3, as well as x and y^2. So, taking out this common factor of $3xy^2$ we get

$$3x^2y^3 + 15xy^4 - 21x^3y^2 = 3xy^2(xy + 5y^2 - 7x^2). \blacksquare$$

Remark: This is the easiest method of factoring and should always be done before going on to the rest.

Exercises 4.1 Factor the following expressions:

1) $2xy + 4x$

2) $6wz + 2wzt$

3) $xy + 4x + 2xw$

4) $6x^2y + 3xy + 9xy^2$

5) $10x^8y^6 + 25x^2y^4 + 20x^3y^{10}$

6) $2\sin^2 x \cos y + \sin x \cos^2 y$

7) $24x^2yz + 2xy^2z^2 + 4xyz^3$

4.2 Special Formulas

(i) $x^2 - y^2 = (x+y)\cdot(x-y)$

(ii) $x^3 + y^3 = (x+y)\cdot(x^2 - xy + y^2)$

(iii) $x^3 - y^3 = (x-y)\cdot(x^2 + xy + y^2)$

(iv) $x^2 + (a+b)x + ab = (x+a)\cdot(x+b)$

(v) $acx^2 + (bc+ad)x + bd = (ax+b)\cdot(cx+d)$

(vi) $x^2 + 2xy + y^2 = (x+y)^2$ and $x^2 - 2xy + y^2 = (x-y)^2$

[The formulas of (vi) are special cases of the two previous formulas. They are called <u>perfect squares</u>.]

<u>Remark:</u>   You should know formulas (i), (ii), (iii), and (vi) "in your sleep." The others, especially (v), are not so critical. All of them can be checked by multiplying out the right-hand sides.

Formula (i) is called the <u>difference of two squares</u> and comes up a lot. Consider the following examples.

Example 1:

a) $z^2 - 9 = (z + 3)\cdot(z - 3)$ (Here we are using (i) with $x = z$ and $y = 3$.)

b) $x^4 - y^2 = (x^2 + y)\cdot(x^2 - y)$ (Now x is replaced by x^2.)

c) $(x-y)^2 - 4y^2 = (x-y+2y)\cdot(x-y-2y)$

$$= (x + y)\cdot(x - 3y)$$

d) $\sin^2 x - 3\cos^2 x = (\sin x + \sqrt{3}\cos x)\cdot(\sin x - \sqrt{3}\cos x)$ ∎

Formula (ii) is called the <u>sum of two cubes</u>, while formula (iii) is the <u>difference of two cubes</u>. Notice how similar they are. One way to remember the signs in the formula is as follows: whatever sign is between the two cubes matches the sign between x and y in the first factor of the product, while the sign in front of the xy in the second factor is opposite to that. [Notice the $-xy$ term in the middle of formula (ii) and the xy term in the middle of formula (iii). It sometimes surprises people.]

Example 2:

    a) $a^3 + 8b^3 = a^3 + (2b)^3 = (a+2b) \cdot (a^2 - 2ab + 4b^2)$

    b) $27x^3 + 64y^3z^6 = (3x)^3 + (4yz^2)^3$

$$= (3x + 4yz^2) \cdot (9x^2 - 12xyz^2 + 16y^2z^4)$$

          Here we are using Formula (ii) with x replaced by $3x$ and y replaced by $4yz^2$.

    c) $\sin^3 8x + 27 \tan^3 x$ is the sum of the cubes of $\sin 8x$ and $3 \tan x$. So,

$$\sin^3 8x + 27 \tan^3 x =$$

$$(\sin 8x + 3 \tan x) \cdot (\sin^2 8x - 3 \sin 8x \, \tan x + 9 \tan^2 x)$$

    d) $a^3 + 2b^3$ is the sum of the cubes of a and $\sqrt[3]{2}\, b$ (which equals $2^{\frac{1}{3}} b$). So,

$$a^3 + 2b^3 = (a + 2^{\frac{1}{3}}b) \cdot (a^2 - 2^{\frac{1}{3}}ab + 2^{\frac{2}{3}}b^2). \blacksquare$$

The usage of Formula (iii) goes pretty much the same way as that of (ii).

Example 3:

$$x^3 - 64y^6 = (x - 4y^2) \cdot (x^2 + 4xy^2 + 16y^4)$$

Again watch the sign of that middle term in the last factor! $\blacksquare$

Formula (iv) is used by noticing that the middle term is the <u>sum</u> of a and b, while the last term is their <u>product</u>.

Example 4:   Factor $x^2 + 5x + 6$.

Solution: We are looking for two numbers a and b, so that their product is 6 and their sum is 5. Well, 6 factors only as 2 times 3, 6 times 1, -2 times -3, or -6 times -1. Of all these possibilities the only pair to add up to 5 is 2 and 3.

So: $x^2 + 5x + 6 = (x+2) \cdot (x+3)$. ∎

Example 5:   Factor $y^2 - 10y + 21$.

Solution: We need two numbers whose product is 21, so we think of 3 and 7, -3 and -7, 21 and 1, and -21 and -1. However, we notice that their sum must be -10, so we use -3 and -7.

$$y^2 - 10y + 21 = (y-3) \cdot (y-7).$$ ∎

Example 6:   Factor $s^2 + 2s - 8$.

Solution: Now we need two numbers whose product is -8 and sum is 2. The number -8 is the product of -8 and 1, -4 and 2, 4 and -2, or 8 and -1. The correct pair that adds up to 2 is 4 and -2, hence

$$s^2 + 2s - 8 = (s+4) \cdot (s-2).$$ ∎

Example 7:   Factor $a^4 - a^2 - 6$.

Solution: This looks different, but if we let $x = a^2$ we get $x^2 - x - 6$, and then this method of factoring applies. Now, -6 is the product of -6 and 1, -3 and 2, -2 and 3, and -1 and 6. Since their sum has to be -1, the only possible choice is -3 and 2. So,

$$x^2 - x - 6 = (x-3) \cdot (x+2)$$

and hence

$$a^4 - a^2 - 6 = (a^2 - 3) \cdot (a^2 + 2).$$

This can be factored further to give

$$a^4 - a^2 - 6 = (a - \sqrt{3})(a + \sqrt{3})(a^2 + 2).$$ ∎

<u>Remark:</u>   In the last example the factorization to $(a^2-3)\cdot(a^2+2)$ is as far as you can go if you factor "over the integers," meaning that all the coefficients must be integers. If you factor "over the reals," you must go the extra step and get $(a-\sqrt{3})(a+\sqrt{3})(a^2+2)$. In some other contexts, but not in calculus, you can even factor a^2+2 "over the complex numbers" into $(a+\sqrt{2}\,i)(a-\sqrt{2}\,i)$. In most situations in calculus, you'll need to factor only over the reals.

Example 8:   Factor $y^2+10y-24$.

Solution: Notice how -24 is the product of many pairs, so this method of factoring can get to be tedious – in fact it can be a real pain! Since their sum is 10, the pair that works is 12 and -2, so

$$y^2+10y-24 = (y-2)\cdot(y+12). \blacksquare$$

<u>Ray of Hope</u>: If this method is too tedious, or too tough, or if the factors are not whole numbers (which happens often), you can use a different method based on the quadratic formula and the factor theorem. For example, x^2-2x-4 can't be factored easily without the factor theorem, which we will see in Section 4.4.

Formula (v) is usually really tough to use because too many possibilities need to be checked, unless you're lucky and hit the right one quickly. Your best bet in this case is to factor out the coefficient of the first term and work with the remaining expression. And there's always the factor theorem.

Exercises 4.2   Factor the following expressions. (Here, to factor means to factor as far as possible over the reals; see Example 7.)

1) $4y^2-9z^2$ 2) $16x^4-y^6$ 3) $8s^3+27t^3$ 4) $2x^3+64y^3$

5) $8s^3-27t^3$ 6) $64z^3-9t^3$ 7) x^2+2x+1 8) x^2+6x+8

9) $x^2-2x-24$ 10) a^4-2a^2-24 11) s^6-7s^3-8 12) $3x^2+x-2$

4.3 Grouping

This method sometimes, but not always, does the job. Here's how it works.

Example 1:   Factor $10xy + 15y + 4x + 6$.

Solution: As written, there are no common factors, but notice that the first and second terms have a common factor of $5y$, while the third and fourth terms have a common factor of 2, giving

$$(10xy + 15y) + (4x + 6) = 5y(2x + 3) + 2(2x + 3).$$

Now these two terms have a common factor of $2x+3$, which we factor out to get

$$10xy + 15y + 4x + 6 = (2x+3)\cdot(5y+2).$$

Notice that the expression is in factored form. ∎

Example 2:   Factor $6ax + 3ay - 4bx - 2by + 10x + 5y$.

Solution: You can group in more than one way. Let's try

$$(6ax + 3ay) + (-4bx - 2by) + (10x + 5y)$$

$$= 3a(2x + y) + (-2b)(2x + y) + 5(2x + y)$$

$$= (2x + y)\cdot(3a - 2b + 5). ∎$$

Remarks: a)   In the above example, we could have grouped differently. Noticing that some terms have an x factor and some a y factor we could have done this:

$$6ax + 3ay - 4by - 2by + 10x + 5y$$

$$= (6ax - 4bx + 10x) + (3ay - 2by + 5y)$$

$$= 2x(3a - 2b + 5) + y(3a - 2b + 5)$$

$$= (3a - 2b + 5)(2x + y)$$

The resulting factorization is the same as before. Always, if grouping is going to work, the resulting factorization will be the same regardless of how you group.

b)   Not all expressions can be factored by grouping; in fact, some expressions cannot be factored <u>by any method</u>. If an expression has a prime number of terms, it definitely can't be factored by <u>grouping</u>. (Can you explain why?)

c)   The resulting factors may need further factoring afterwards.

Exercises 4.3 Factor the following completely:

1) $3ax + 2ay + 3bx + 2by$

2) $x^4 - x^3y + x - y$

3) $x^{10} + x^6y^2 + x^4y^3 + y^5$

4) $6x^3y - 4xy^3 + 12yx^2 - 8y^3$

5) $3x^2 + 5xy + 7x + 3xy + 5y^2 + 7y$

4.4 The Factor Theorem and Long Division

Here it comes, the long-awaited entrance of the ultimate in theoretical sensations. Yes folks, the one, the only, the celebrated factor theorem. Here is the pearl of wisdom. (Let it rip Frank!)

Factor Theorem: Let $P(x)$ be a polynomial. Then $x-a$ (for some number a) is a factor of $P(x)$ iff $P(a) = 0$. ("iff" means: if and only if.)

Remark: So what's the big deal with this factor theorem? Here's what. Suppose you want to factor a polynomial $P(x)$, and the other methods don't look promising. But suppose you can find some number a such that $P(a) = 0$. [You can do this by trying easy numbers like $a = \pm 1, \pm 2$, etc. to see if you are lucky and find such an a. If you have a graphing calculator, you can see where the graph of the function crosses or touches the x-axis. By zooming in on one such point, you may obtain an approximation for such a number a where $P(a) = 0$.] IF YOU CAN FIND an a with $P(a) = 0$, then you are sure, because of the factor theorem, that $x - a$ is a factor of $P(x)$. So $P(x) = (x - a) \cdot (\text{something})$. That "something" can be found by long division, and so you've factored $P(x)$!

Example 1:   Factor $P(x) = x^3 - 2x^2 - 5x + 6$.

Solution: This one looks tough. The earlier methods of factoring would give you nothing. Zip. Zero. Nada. Goose eggs. But notice that if you put $x = 1$, you get $1^3 - 2 \cdot 1^2 - 5 \cdot 1 + 6 = 0$. Aha! So the F. T. (factor theorem) says that $x - 1$ is a factor. So:

$$P(x) = (x-1) \cdot (\text{something}).$$

You're half done. To get the other factor, use long division:

$$x-1 \overline{)x^3 - 2x^2 - 5x + 6}$$ (Notice that both expressions are in descending powers!)

Divide the first term into the first term (x into x^3 gives x^2). Put the x^2 on top, multiply it by the entire divisor $x - 1$, and subtract it from the dividend $x^3 - 2x^2 - 5x + 6$:

$$
\begin{array}{r}
x^2 \\
x-1 \overline{)x^3 - 2x^2 - 5x + 6} \\
\underline{x^3 - x^2 } \\
-x^2 - 5x + 6
\end{array}
$$

(The easiest way to subtract $x^3 - x^2$ is to <u>mentally</u> change the sign of $x^3 - x^2$, getting $-x^3 + x^2$, and <u>adding</u>.) Continue in the same way: x into $-x^2$ gives $-x$, etc., to get:

$$
\begin{array}{r}
x^2 - x \phantom{{}- 5x + 6} \\
x-1 \overline{)\,x^3 - 2x^2 - 5x + 6\,} \\
\underline{x^3 - x^2 \phantom{{}- 5x + 6}} \\
-x^2 - 5x + 6 \\
\underline{-x^2 + x \phantom{{}+ 6}} \\
-6x + 6
\end{array}
$$

One more time: x into $-6x$ gives -6. So:

$$
\begin{array}{r}
x^2 - x - 6 \phantom{{}+ 6} \\
x-1 \overline{)\,x^3 - 2x^2 - 5x + 6\,} \\
\underline{x^3 - x^2 \phantom{{}- 5x + 6}} \\
-x^2 - 5x + 6 \\
\underline{-x^2 + x \phantom{{}+ 6}} \\
-6x + 6 \\
\underline{-6x + 6} \\
0
\end{array}
$$

We knew that the remainder would be 0 because the F. T. told us so.

Hence $P(x) = (x-1) \cdot (x^2 - x - 6)$.

So $P(x)$ is factored, but not factored completely as yet because $(x^2 - x - 6)$ can be factored further to give

$$P(x) = (x-1) \cdot (x - 3) \cdot (x + 2).$$

Success ! ∎

Example 2: Factor $2x^2 + 3x - 2$.

Solution: You could try special formula (v), but it would take too long. Remembering the F. T., all you need is a number "a" that makes $2a^2 + 3a - 2 = 0$. In ordinary English: you need a solution to $2x^2 + 3x - 2 = 0$. But that's easy, if you recall the quadratic formula,

$$x = \frac{-b \pm \sqrt{b^2 - 4ac}}{2a},$$

where in this context, $a = 2$, $b = 3$, and $c = -2$. Using this formula, you get

$$x = \frac{-3 \pm \sqrt{9 + 16}}{4} = \frac{-3 \pm 5}{4} = \frac{1}{2} \text{ or } -2.$$ So the factor theorem gives us both factors $\left(x - \frac{1}{2}\right)$ and $(x + 2)$, and so

$$2x^2 + 3x - 2 = \left(x - \frac{1}{2}\right) \cdot (x + 2) \cdot (\text{something}).$$

Since the coefficient of the x^2 term on the left is 2, that something <u>must be 2</u>, for equality. So:

$$2x^2 + 3x - 2 = \left(x - \frac{1}{2}\right)(x + 2)\,(2) = (2x - 1) \cdot (x + 2). \ \blacksquare$$

Example 3: Can $x^2 + x + 1$ be factored?

Solution: The F. T. tells us to check for solutions of the equation $x^2 + x + 1 = 0$. The quadratic formula gives

$$x = \frac{-1 \pm \sqrt{1 - 4}}{2},$$

which is not a real number, because negative numbers don't have square roots among the real numbers. (See Section 1.5 for further explanation of this.) So there is no real number a to make $a^2 + a + 1$ equal to 0. Hence the F. T. says that $x^2 + x + 1$ <u>can't be factored</u>. (Such quadratics are called <u>irreducible</u>.) $\blacksquare$

<u>To sum up:</u>   You can always use the F. T. to factor quadratics by using the quadratic formula, but for other polynomials $P(x)$, you'll need to be lucky to find a number a that makes $P(a) = 0$. Try $a = \pm 1, \pm 2$, etc.

In real-life problems, scientists and engineers use calculators or computers to find, or approximate, values for a where $P(a) = 0$.

Exercises 4.4 Factor the following expressions, if possible:

1) $x^3 - 3x + 2$ 2) $2x^3 + x + 3$ 3) $x^3 - x + 6$

4) $2x^2 - 3x + 4$ 5) $x^2 - 3x - 2$ 6) $24x^2 - 48x - 72$

4.5 Rationalizing Numerators or Denominators using Conjugates

Consider the expression $\dfrac{x - \sqrt{2}}{5}$. Notice that the numerator has two terms, one of which is a square root. For various reasons, you may not want a square root in the numerator. You can always get rid of it by multiplying and dividing by its conjugate $x + \sqrt{2}$. (You find the conjugate by changing the sign between the two terms.) So you will get

$$\frac{x - \sqrt{2}}{5} = \frac{x - \sqrt{2}}{5} \cdot \frac{x + \sqrt{2}}{x + \sqrt{2}} = \frac{(x - \sqrt{2}) \cdot (x + \sqrt{2})}{5 \cdot (x + \sqrt{2})}$$

$$= \frac{x^2 - 2}{5 \cdot (x + \sqrt{2})} .$$

Remarks: a) The top is now root-free.

b) The little devil's conjugate has popped up in the denominator instead. Depending on the problem you're solving, that may not cause any difficulties.

c) It all boils down to this: you can exchange a square root in the top for a square root in the bottom, or vice versa, whichever is better in a particular case.

Example 1:   Rationalize the denominator of $\dfrac{x^2 - 3}{x + \sqrt{3}}$.

Solution: This means get rid of the root in the bottom. So multiply both top and bottom by the conjugate $x - \sqrt{3}$, giving

$$\frac{x^2 - 3}{x + \sqrt{3}} = \frac{x^2 - 3}{x + \sqrt{3}} \cdot \frac{x - \sqrt{3}}{x - \sqrt{3}} = \frac{(x^2 - 3)(x - \sqrt{3})}{(x + \sqrt{3})(x - \sqrt{3})}$$

$$= \frac{(x^2 - 3)(x - \sqrt{3})}{(x^2 - 3)} = (x - \sqrt{3}) . \blacksquare$$

Remark: When you're working with a fractional expression, and either the top or bottom has two terms, one (or both) of which is a square root, it is often useful to rationalize it. This comes up in limits, but also in many other cases. If you have such an expression, it <u>should always pop into your mind that one option is to rationalize</u>, which may help.

Example 2:   Rationalize $\dfrac{x^4 - 25}{x - \sqrt{5}}$.

Solution: Multiply both top and bottom by the conjugate $x + \sqrt{5}$, giving

$$\frac{x^4 - 25}{x - \sqrt{5}} = \frac{x^4 - 25}{x - \sqrt{5}} \cdot \frac{x + \sqrt{5}}{x + \sqrt{5}} = \frac{(x^4 - 25)(x + \sqrt{5})}{(x^2 - 5)},$$

which can be factored to

$$= \frac{(x^2 - 5)(x^2 + 5)(x + \sqrt{5})}{(x^2 - 5)} = (x^2 + 5)(x + \sqrt{5}),$$

which is probably easier to deal with. ■

<u>Remark:</u>   You may have noticed that you can get the same result by factoring the top completely and then canceling:

$$\frac{x^4 - 25}{x - \sqrt{5}} = \frac{(x^2 + 5)(x^2 - 5)}{x - \sqrt{5}} = \frac{(x^2 + 5)(x + \sqrt{5})(x - \sqrt{5})}{(x - \sqrt{5})} = (x^2 + 5)(x + \sqrt{5}).$$

That's true, but you've still got to know how to rationalize (see exercises 5.1).

Example 3:   Rationalize $\dfrac{\sqrt{x+h} - \sqrt{x}}{h}$.

Solution: Multiply both top and bottom by the conjugate $\sqrt{x+h} + \sqrt{x}$, giving

$$\frac{\sqrt{x+h} - \sqrt{x}}{h} = \frac{\left(\sqrt{x+h} - \sqrt{x}\right)\left(\sqrt{x+h} + \sqrt{x}\right)}{h\left(\sqrt{x+h} + \sqrt{x}\right)},$$

which when simplified gives

$$\frac{\sqrt{x+h} - \sqrt{x}}{h} = \frac{x+h-x}{h\left(\sqrt{x+h} + \sqrt{x}\right)} = \frac{1}{\left(\sqrt{x+h} + \sqrt{x}\right)}. \quad \blacksquare$$

The following example presents part of the calculation needed to find the derivative of $f(x) = \dfrac{1}{\sqrt{x}}$.

Example 4: Write $\dfrac{1}{\sqrt{x+h}} - \dfrac{1}{\sqrt{x}}$ as one fraction, and rationalize the resulting numerator.

Solution: First, use a common denominator to get

$$\frac{\sqrt{x} - \sqrt{x+h}}{\sqrt{x+h} \ \sqrt{x}}.$$

Now multiply both top and bottom by the conjugate $\sqrt{x} + \sqrt{x+h}$.

$$\frac{\sqrt{x} - \sqrt{x+h}}{\sqrt{x+h} \ \sqrt{x}} = \frac{\sqrt{x} - \sqrt{x+h}}{\sqrt{x+h} \ \sqrt{x}} \cdot \frac{\sqrt{x} + \sqrt{x+h}}{\sqrt{x} + \sqrt{x+h}}$$

$$= \frac{x - (x+h)}{\sqrt{x+h} \ \sqrt{x}\left(\sqrt{x} + \sqrt{x+h}\right)}$$

$$= \frac{-h}{\sqrt{x+h} \ \sqrt{x}\left(\sqrt{x} + \sqrt{x+h}\right)}. \quad \blacksquare$$

Exercises 4.5 Rationalize the top or bottom, and simplify:

1) $\dfrac{7}{\sqrt{2} - 1}$ 2) $\dfrac{x + \sqrt{2}}{2}$ 3) $\dfrac{3}{x - \sqrt{7}}$ 4) $\dfrac{x + 1}{x + \sqrt{11}}$

5) $\dfrac{x^2 - 3}{x - \sqrt{3}}$ 6) $\dfrac{x^4 - 36}{x + \sqrt{6}}$ 7) $\dfrac{x^8 - 9}{x^2 + \sqrt{3}}$

8) Let $f(x) = \dfrac{1}{\sqrt{2x}}$. Calculate $\dfrac{f(x+h) - f(x)}{h}$ and simplify as in Example 4.

4.6 Extracting Factors from Radicals

Radicals can be difficult when computing things, so it usually pays to make them as simple as possible. Extracting factors from under the radical sign is one way of simplifying.

Example 1:   Simplify $\sqrt[3]{250x^4y^3}$.

Solution: Since $(ab)^n = a^n \cdot b^n$ for all n (as long as both sides are defined!), we have $(ab)^{\frac{1}{3}} = a^{\frac{1}{3}} \cdot b^{\frac{1}{3}}$ and $\sqrt[3]{a \cdot b} = \sqrt[3]{a} \cdot \sqrt[3]{b}$. In this example you can write $250x^4 y^3 = (125x^3y^3)(2x)$. (We chose the first factor to be a __perfect cube__.)

So $\sqrt[3]{250x^4y^3} = \sqrt[3]{125x^3y^3}\sqrt[3]{2x} = 5xy\sqrt[3]{2x}$. ■

Example 2:   Simplify the radical by extracting all that you can from $\sqrt{25x^8\sin x}$.

Solution: $\sqrt{25x^8 \sin x} = \sqrt{(25x^8)(\sin x)}$

$$= \sqrt{25x^8}\sqrt{\sin x} ,$$

$$= 5x^4\sqrt{\sin x} . ■$$

<u>Remarks:</u> a) Notice that in Example 1 as you "pulled" the factor $125x^3y^3$ out of the radical, it "became" its cube root $5xy$. Similarly, as you "pulled" $25x^8$ out of the root in Example 2, it "changed into" its square root $5x^4$.

   b) When working with square roots and other even-powered roots, you must remember that $\sqrt{a^2} = |a|$, not just a . The next example will illustrate.

Example 3:   Simplify $\sqrt{36xy^2z^3}$.

Solution: $\sqrt{36xy^2z^3} = \sqrt{(36y^2z^2)(xz)}$, where the first factor is a perfect square.

So,

$$\sqrt{36\,x\,y^2 z^3} = \sqrt{(36\,y^2 z^2)}\sqrt{(x\,z)}$$

$$= (\sqrt{36}\sqrt{y^2}\sqrt{z^2})\sqrt{x\,z}$$

$$= 6|y||z|\sqrt{x\,z}. \quad \blacksquare$$

Example 4:   Simplify $f(x) = \sqrt{\sec^2 x + \sec^4 x}$.

Solution: $f(x) = \sqrt{(\sec^2 x)(1 + \sec^2 x)}$,

$$= \sqrt{\sec^2 x}\sqrt{1 + \sec^2 x} ,$$

$$= |\sec x|\sqrt{1 + \sec^2 x} . \quad \blacksquare$$

Example 5:   Simplify $f(x) = \sqrt[4]{16 x^8 \cos x}$.

Solution: $f(x) = \sqrt[4]{(16 x^8)(\cos x)}$,

$$= \sqrt[4]{16 x^8}\sqrt[4]{\cos x} ,$$

$$= \sqrt[4]{16}\,\sqrt[4]{x^8}\,\sqrt[4]{\cos x} ,$$

$$= 2\,|x^2|\sqrt[4]{\cos x} .$$

Here, the stuff in the absolute value sign, x^2 , is always non-negative, and so $|x^2| = x^2$. We can thus say

$$f(x) = 2 x^2 \sqrt[4]{\cos x} . \quad \blacksquare$$

Example 6:   Simplify $g(x) = \sqrt[5]{32 x^{10} \tan^2 x}$.

Solution: $g(x) = \sqrt[5]{32 x^{10}}\sqrt[5]{\tan^2 x} = 2 x^2 \sqrt[5]{\tan^2 x} . \quad \blacksquare$

Example 7:   Simplify $\sqrt{x^2 y^6 + 3x^5 y^4}$.

Solution: The stuff in the radical is not factored yet, so you must do that before you can extract any factors!

$$\sqrt{x^2 y^6 + 3x^5 y^4} = \sqrt{x^2 y^4 (y^2 + 3x^3)}$$

$$= \sqrt{x^2 y^4} \sqrt{y^2 + 3x^3}$$

$$= |x||y^2| \sqrt{y^2 + 3x^3}$$

$$= |x| y^2 \sqrt{y^2 + 3x^3} . \quad \blacksquare$$

Exercises 4.6 Extract as much as you can from the following roots:

1) $\sqrt{16x^2}$

2) $\sqrt{4x^2 + 8x^4}$

3) $\sqrt[3]{54x^4}$

4) $\sqrt{3x^{12} y}$

5) $\sqrt{5x^4 + 3x^8}$

6) $\sqrt[3]{27x^6 \cos x}$

7) $\sqrt{8\pi^2 x^3 y^4}$

8) $\sqrt[4]{x^5 y^4 + x^6 y^{10}}$

Chapter 5

Rational Operations and Expressions

Recall from Chapter 1 how to add fractions: $\dfrac{a}{b} + \dfrac{c}{d} = \dfrac{ad + bc}{bd}$. First form a common denominator, namely bd, then change $\dfrac{a}{b}$ to $\dfrac{ad}{bd}$, and $\dfrac{c}{d}$ to $\dfrac{bc}{bd}$, at which point you can add them to get the result. The same method applies, no matter how tough these expressions look.

Example 1: Simplify $\dfrac{1}{x} - \dfrac{1}{x-1}$.

Solution:
$$\frac{1}{x} - \frac{1}{x-1} = \frac{1 \cdot (x-1) - 1 \cdot x}{x(x-1)}$$

$$= \frac{x - 1 - x}{x(x-1)} = \frac{-1}{x(x-1)}. \quad \blacksquare$$

Example 2: Simplify $\dfrac{3x + y}{x + y} + \dfrac{x - y}{x + 2y}$.

Solution: Finding a common denominator and adding gives

$$\frac{(3x + y)(x + 2y) + (x + y)(x - y)}{(x + y)(x + 2y)}$$

This needs a little "cleaning up." Let's multiply out the top, to get

$$\frac{3x^2 + xy + 6xy + 2y^2 + x^2 + xy - xy - y^2}{(x + y)(x + 2y)}$$

$$= \frac{4x^2 + 7xy + y^2}{(x + y)(x + 2y)}.$$

For most purposes, you would not want to multiply out the denominator. $\quad \blacksquare$

Example 3:   Simplify $\dfrac{\dfrac{1}{t+1} - \dfrac{1}{t}}{\dfrac{1}{t+1} + \dfrac{1}{t}}$.

Solution: Working separately with the numerator and denominator,

$$\frac{1}{t+1} - \frac{1}{t} = \frac{t - (t+1)}{t(t+1)} = \frac{-1}{t(t+1)}$$

$$\frac{1}{t+1} + \frac{1}{t} = \frac{t + (t+1)}{t(t+1)} = \frac{2t+1}{t(t+1)}$$

So:

$$\frac{\dfrac{1}{t+1} - \dfrac{1}{t}}{\dfrac{1}{t+1} + \dfrac{1}{t}} = \frac{\dfrac{-1}{t(t+1)}}{\dfrac{2t+1}{t(t+1)}} = \frac{-1}{t(t+1)} \cdot \frac{t(t+1)}{2t+1} = \frac{-1}{2t+1}.$$

This answer is as simple as it gets.   ■

Example 4:   Simplify $\dfrac{x^{-1} + y^{-1}}{(xy)^{-1}}$.

Solution: First convert the expression to fractions:

$$\frac{x^{-1} + y^{-1}}{(xy)^{-1}} = \frac{\dfrac{1}{x} + \dfrac{1}{y}}{\dfrac{1}{xy}}$$

$$= \frac{\dfrac{1 \cdot y + 1 \cdot x}{xy}}{\dfrac{1}{xy}} = \frac{\dfrac{y+x}{xy}}{\dfrac{1}{xy}} = \frac{y+x}{1} = y + x. \quad ■$$

Example 5: Simplify $\dfrac{\dfrac{1}{x+h}-\dfrac{1}{x}}{h}$.

Solution:

$$\frac{\dfrac{1}{x+h}-\dfrac{1}{x}}{h}=\frac{\dfrac{x-(x+h)}{x(x+h)}}{h}$$

$$=\frac{\dfrac{x-x-h}{x(x+h)}}{h}=\frac{\dfrac{-h}{x(x+h)}}{h}\ .$$

Notice that both the top and bottom have a factor of h, which can be canceled to obtain

$$\frac{h\left(\dfrac{-1}{x(x+h)}\right)}{h}=\frac{-1}{x(x+h)}\ . \quad\blacksquare$$

Example 6: Simplify $\dfrac{3x^2y^5}{24\,x\,y^2}$.

Solution: Cancel factors that are common to the numerator and the denominator:

$$\frac{3x^2y^5}{24\,x\,y^2}=\frac{x\,y^3}{8}\ . \quad\blacksquare$$

Example 7: Simplify $\dfrac{5a^2b^3-15ab}{100\,a^2\,b^2}$.

Solution: Remember to avoid "creative canceling." Cancel only those things that are factors of both the <u>entire top</u> and the <u>entire bottom</u>. Factor the numerator to obtain

$$\frac{5a^2b^3-15ab}{100\,a^2\,b^2}=\frac{5ab(ab^2-3)}{100\,a^2\,b^2}$$

$$= \frac{a b^2 - 3}{20 a b}. \quad \blacksquare$$

Example 8:   Simplify $\dfrac{27 x^2 \left(\dfrac{y}{z}\right)^{-5}}{(3 x y z^4)^2}$.

Solution: $\dfrac{27 x^2 \left(\dfrac{y}{z}\right)^{-5}}{(3 x y z^4)^2} = \dfrac{27 x^2 \left(\dfrac{z}{y}\right)^{5}}{9 x^2 y^2 z^8} = \dfrac{27 x^2 y^{-5} z^5}{9 x^2 y^2 z^8}$

$$= 3 y^{-7} z^{-3} = \frac{3}{y^7 z^3}. \quad \blacksquare$$

Example 9:   Let $f(x) = x^2 + \dfrac{1}{x}$.

Calculate: a) $f(x + \Delta x)$

b) $\dfrac{f(x + \Delta x) - f(x)}{\Delta x}$ (You'll see lots of these !)

Note: Δx is to be interpreted as just another variable that could just as well be called h (which it often is), or y or w or whatever.

Solution: a) $f(x + \Delta x) = (x + \Delta x)^2 + \dfrac{1}{x + \Delta x}$.

b) $\dfrac{f(x + \Delta x) - f(x)}{\Delta x} = \dfrac{(x + \Delta x)^2 + \dfrac{1}{x + \Delta x} - \left(x^2 + \dfrac{1}{x}\right)}{\Delta x}$

$$= \frac{x^2 + 2x \cdot \Delta x + (\Delta x)^2 + \dfrac{1}{x + \Delta x} - x^2 - \dfrac{1}{x}}{\Delta x}.$$

Cancel the x^2 's, and subtract the two fractions on top. So

$$\frac{f(x + \Delta x) - f(x)}{\Delta x} = \frac{2x \cdot \Delta x + (\Delta x)^2 + \dfrac{x - (x + \Delta x)}{x(x + \Delta x)}}{\Delta x}$$

$$= \frac{2x \cdot \Delta x + (\Delta x)^2 - \dfrac{\Delta x}{x(x + \Delta x)}}{\Delta x}$$

$$= \frac{\Delta x \left(2x + \Delta x - \dfrac{1}{x(x + \Delta x)} \right)}{\Delta x}$$

$$= 2x + \Delta x - \frac{1}{x(x + \Delta x)} . \qquad \blacksquare$$

Exercises 5.1 Simplify the following expressions:

1) $\dfrac{1}{x - 1} - \dfrac{1}{x}$

2) $\dfrac{x}{x - 1} - \dfrac{2}{x}$

3) $\dfrac{x}{x - 1} - \dfrac{x}{x + 1}$

4) $\dfrac{\dfrac{s + 1}{s - 1} + \dfrac{s - 1}{s + 1}}{\dfrac{1}{s^2 - 1}}$

5) $\dfrac{2(x + h) + 1 - (2x + 1)}{h}$

6) $\dfrac{\dfrac{1}{(x + h)^2} - \dfrac{1}{x^2}}{h}$

7) $x + \dfrac{1}{x + \dfrac{1}{x + \dfrac{1}{x}}}$

8) $\dfrac{\sqrt{2x + 2h} - \sqrt{2x}}{h}$ (Hint: Rationalize your way out of this one!)

9) $\dfrac{3a^2 b - 27ab^2}{(15a^3 b^4)^2}$

10)   Let $f(x) = x^3 + 2x$. Calculate $\dfrac{f(x + h) - f(x)}{h}$.

11) Let $g(x) = \sqrt{x - 3}$. Calculate $\dfrac{g(x + \Delta x) - g(x)}{\Delta x}$ (Hint: Rationalize your way to success!)

Chapter 6

Cyclic Phenomena: The Six Basic Trigonometric Functions

6.1 Angles

The size of an angle can be expressed in several ways. You've used "degrees" for years. They are simple. You've also used "revolutions" when talking about the RPM of an engine. In calculus, "radians" are used as a unit of measure because the rules of calculus are easiest that way. How big is a radian? Here's how big: <u>it's the angle corresponding to an arc length of 1 in a unit circle.</u> Look at the diagram below. A "unit circle" indicates that the radius = 1, and we'll always put the center at (0,0) for convenience. The angle θ as drawn is 1 radian, because the arc length "subtended" (cut off) by the angle has length = 1. By the way, θ is the Greek letter "theta" and is often used for angles.

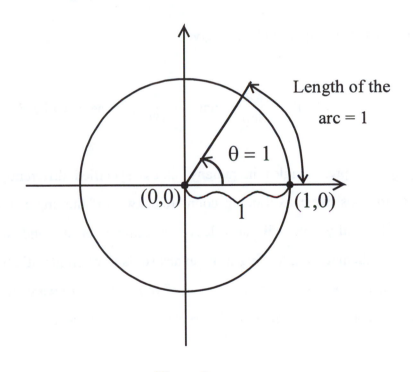

Figure 1

Notice that if you double the length of the arc, you double the angle. In fact, in a unit circle, the arc length equals the angle (if you use radians). Since most of you are probably more comfortable using degrees, we must deal with the key question: how do you convert from degrees to radians, and vice versa? Here's the secret:

One revolution = the complete angle at the center in radians, so

      = the length of the complete arc

      = the circumference of the unit circle

      = $2\pi r$, but $r = 1$, so

      = 2π (radians, that is)

You recall that there are 360 degrees in one full revolution, so

$$\boxed{360^\circ = 2\pi \text{ radians}}.$$

If you divide both sides of this equation by 2π, you obtain

$$1 \text{ radian} = \frac{360^\circ}{2\pi} \cong 57^\circ,$$

while dividing both sides of the equation by 360 gives

$$1^\circ = \frac{2\pi}{360} \text{ radians} = \frac{\pi}{180} \text{ radians} \cong .017 \text{ radians}.$$

  We will <u>always</u> express angles in <u>radians</u>, unless specified differently. Moreover, angles are measured starting at the positive x-axis going counterclockwise if the angle is positive, clockwise if the angle is negative. Typically, we will also leave π simply as π, and will not use any decimal approximation. To a mathematician, leaving π in an answer is fine. In fact it's better than fine because it's exact. Anything else would be less exact. [However, if you go to a hardware store and ask the clerk for π square inches of sheet metal, you will get (and deserve!) some funny stares.]

Example 1:   Convert 27° to radians.

Solution: $360^\circ = 2\pi$ (If things are clear, we usually drop the word "radians.")

$$1^\circ = \frac{2\pi}{360}, \text{ so } 27^\circ = 27 \cdot \frac{2\pi}{360} = \frac{3}{40} \cdot 2\pi = \frac{3\pi}{20}. \quad \blacksquare$$

Example 2:   Convert $-\dfrac{5\pi}{6}$ radians to degrees, and diagram it.

Solution: $2\pi = 360°$

$\pi = 180°$

$$-\frac{5}{6}\cdot\pi = -\frac{5}{6}\cdot 180° = -150°.$$

The angle is measured clockwise, because it is negative, and is shown in the figure below.

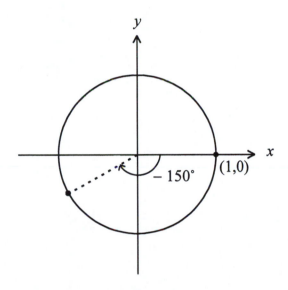

You should commit to memory the following conversion chart since these angles will come up again and again.

Degrees	360	180	90	60	45	30
Radians	2π	π	$\pi/2$	$\pi/3$	$\pi/4$	$\pi/6$

Exercises 6.1

1)   Change to radian measure:

a) $120°$ b) $270°$ c) $135°$ d) $210°$ e) $-150°$ f) $450°$

2)   Change to degrees:

a) $\dfrac{3\pi}{4}$ b) $\dfrac{11\pi}{6}$ c) $-\dfrac{\pi}{3}$ d) 3π e) $\dfrac{9\pi}{2}$ f) $\dfrac{9\pi}{4}$

3)   Draw a large circle and mark each of the angles in exercises 1 and 2.

6.2 Definition of Sin θ and Cos θ

Consider the unit circle, centered at the origin, with an angle of θ radians, as shown in Figure 2.

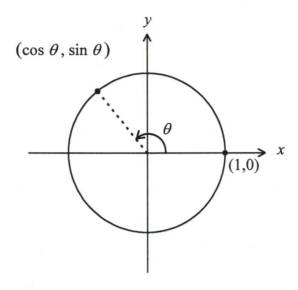

Figure 2

(Notice that the angle is measured from the positive x-axis, counterclockwise.) The dotted line defining the terminal side (end) of the angle θ intersects the circle at a point. As the angle θ changes, so do the coordinates of that point, so each of the coordinates is a function of the angle θ. The first coordinate is called cos θ; the second is called sin θ. These functions are read as "cosine" and "sine" respectively. (See Chapter 9 for a different way of defining them.)

Remarks: a)   Since this point is on the unit circle, its coordinates must satisfy the equation of that circle:

$x^2 + y^2 = 1$, that is $(\cos\theta)^2 + (\sin\theta)^2 = 1$.

b)   To avoid the constant use of brackets, we write $\cos^n\theta$ to mean $(\cos\theta)^n$; similarly for sine, we write $\sin^n\theta$ to mean $(\sin\theta)^n$. Thus $\cos^2\theta + \sin^2\theta = 1$.

c)   Imagine θ going through values from 0 to 2π. Then the point on the circle goes from (1,0) counterclockwise around the circle. Thus, its second coordinate, $\sin\theta$, which is the altitude of the point, goes first from 0 up to 1, then down to -1, then back up to 0, as shown in the graph below.

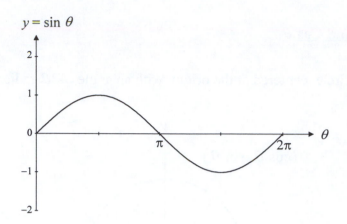

Meanwhile, the first coordinate, cos θ, goes from 1 down to -1, and back to 1.

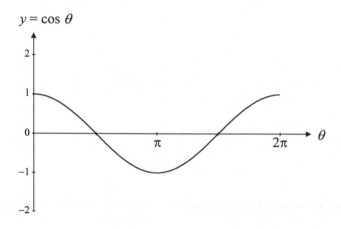

d)   Since θ and $\theta + 2\pi$ correspond to the same point on the unit circle, we have $\sin \theta = \sin(\theta + 2\pi)$. In fact, $\sin \theta = \sin(\theta + 2k\pi)$ for any whole number k. We say that $\sin \theta$ is <u>periodic</u> with period 2π. The same is true of $\cos \theta$. Remember that $2\pi k$ radians, where k is a whole number, is exactly k revolutions.

e)   We just happened to use the variable name θ. We could also use $x, y, z, w, u, v,$ ☺, ☹, or any other letter or symbol desired! To be consistent with the function notation $f(x)$ we've come to know and love, we will use the variable x. So if we allow x to be any real number, we get the <u>full</u> graph of $f(x) = \sin x$ and $g(x) = \cos x$.

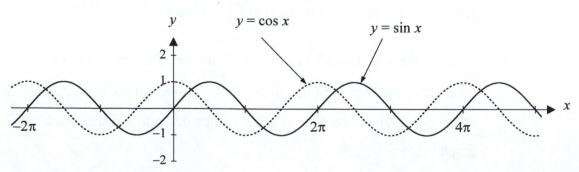

f)   You might guess that sin x and cos x and other trigonometric functions would be useful for studying waves, daily temperature averages through the seasons, alternating current, electronic signals, and lots of other cyclic phenomena. You'd be perfectly right. These functions are the basic tools of the trade.

Example 1:   What is $\cos\dfrac{3\pi}{2}$ and $\sin\dfrac{3\pi}{2}$?

Solution: To determine the sine and cosine of some angles is really quite easy. Just remember the diagram below. The angle $^{3\pi}/_2$ is 3/4 of the way around the circle, ending up at the point $(0,-1)$. For angles that end up on the coordinate axes, the sine and cosine are given immediately because we know exactly what the coordinates of those points are.

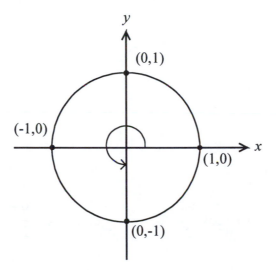

Beginning with the point $(1,0)$, and traveling counterclockwise, the four points labeled in the figure above are the endpoints for the angles 0, $^{\pi}/_2$, π, $^{3\pi}/_2$, and 2π. In this case, for $^{3\pi}/_2$ the endpoint has an x-coordinate of 0, and a y-coordinate of -1. Just remember $\underline{x}$ before $\underline{y}$, **c** before **s**. The x-coordinate is the **c**osine, and the y-coordinate is **s**ine. So

$$\cos\frac{3\pi}{2}=0 \ \text{ and } \ \sin\frac{3\pi}{2}=-1. \ \blacksquare$$

Example 2:   What is $\cos \dfrac{9\pi}{2}$?

Solution: First of all, $\dfrac{9\pi}{2} = 4\pi + \dfrac{\pi}{2}$, and 4π radians equals two complete

revolutions. So the angle $4\pi + \dfrac{\pi}{2}$ goes twice around the circle (which does

nothing), and then another $90°$, ending up at the top of the circle, at the point

$(0,1)$. The cosine is the first coordinate of that point. So

$$\cos \dfrac{9\pi}{2} = 0. \blacksquare$$

Exercises 6.2

1) Evaluate:

a) $\sin \dfrac{7\pi}{2}$ b) $\cos \dfrac{5\pi}{2}$ c) $\cos \dfrac{-9\pi}{2}$ d) $\sin 101\pi$

2) Assuming k is a whole number, evaluate the following:

a) $\sin(\pi/2 + 2k\pi)$ b) $\cos(-\pi/2 + 2k\pi)$ c) $\sin k\pi$ d) $\cos k\pi$

6.3 The Other Trigonometric Functions

Having defined the functions $f(x) = \sin x$ and $f(x) = \cos x$ for any real number x, there are four more basic trigonometric functions that you will encounter on your travels:

$$\tan x = \frac{\sin x}{\cos x} \quad \text{(called "tangent")}$$

$$\cot x = \frac{\cos x}{\sin x} \quad \text{(called "cotangent")}$$

$$\sec x = \frac{1}{\cos x} \quad \text{(called "secant")}$$

$$\csc x = \frac{1}{\sin x} \quad \text{(called "cosecant")}$$

<u>Remark:</u> At certain values of x, the denominator is zero; hence at those values of x, that function is not defined. For example, $\tan x$ is not defined at $x = \frac{\pi}{2}$, because $\cos \frac{\pi}{2} = 0$.

Exercises 6.3

1) Evaluate $\tan \frac{5\pi}{2}$, $\csc \frac{5\pi}{2}$, $\sec \frac{5\pi}{2}$, and $\cot \frac{5\pi}{2}$.

2) Evaluate $\tan \frac{-\pi}{2}$, $\csc \frac{-\pi}{2}$, $\sec \frac{-\pi}{2}$, and $\cot \frac{-\pi}{2}$.

3) Notice that $\frac{1}{x}$ gets large if x gets small – for example, $\frac{1}{\frac{1}{1000}} = 1000$. Also recall that $\csc x = \frac{1}{\sin x}$. Thus if we draw the graph of $\sin x$ from 0 to $\frac{\pi}{2}$, we can envision the graph of $\csc x$ as shown below.

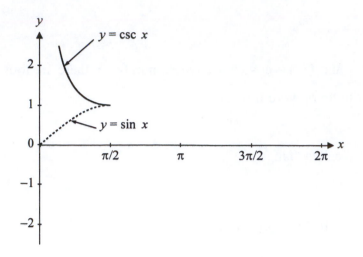

Complete this sketch for x in $[0, 2\pi]$ – i.e., graph csc x on $[0, 2\pi]$ (except at certain points where it is not defined).

4) Similarly graph sec x on $[0, 2\pi]$, and then for several periods (except at certain ...) .

5) Graph tan x on $[-\pi, \pi]$ (except at certain ...) .

6) Graph cot x on $[0, 2\pi]$ (except at certain ...) .

7) Using the unit circle, estimate $\cos(-1.2)$. (A rough approximation is fine.)

(Hint: $\dfrac{\pi}{2} \approx \dfrac{3}{2} = 1.5$, so 1.2 is an angle a little smaller than $90°$.)

8) Estimate sin (3.0).

6.4 Basic Identities

We know that

$$\boxed{\sin^2 x + \cos^2 x = 1}.$$

Dividing both sides of this equation by $\sin^2 x$, we get

$$\frac{\sin^2 x}{\sin^2 x} + \frac{\cos^2 x}{\sin^2 x} = \frac{1}{\sin^2 x}$$

That is,
$$\boxed{1 + \cot^2 x = \csc^2 x}.$$

Dividing the top equation by $\cos^2 x$, we get

$$\boxed{\tan^2 x + 1 = \sec^2 x}.$$

Of course, you can play with these identities to make them look a little different. Take the first one for instance:

$$\sin^2 x = 1 - \cos^2 x \quad \text{and} \quad \cos^2 x = 1 - \sin^2 x ,$$

so that
$$\sin x = \pm \sqrt{1 - \cos^2 x} \quad \text{and} \quad \cos x = \pm \sqrt{1 - \sin^2 x} .$$

Do this with the other two. You will need the results!

Exercises 6.4

1)   Draw the unit circle, centered at the origin.

    a)   For some number θ, draw the angles θ and $-\theta$.

    b)   Write $\cos(-\theta)$ in terms of $\cos\theta$.

    c)   Write $\sin(-\theta)$ in terms of $\sin\theta$.

    d)   What does that tell you about $\tan(-\theta)$ in terms of $\tan\theta$?

2)   a)   Express $\tan\theta$ in terms of $\sec\theta$, and vice versa.

    b)   Express $\cot\theta$ in terms of $\csc\theta$, and vice versa.

6.5 <u>Special Angles</u> $\left(\frac{\pi}{4}, \frac{\pi}{6}, \frac{\pi}{3}\right)$

Example 1:   Find $\sin\dfrac{\pi}{4}$.

Solution: Since $\frac{\pi}{4} = 45°$, the picture looks like

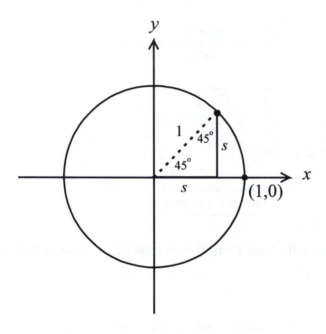

where s is some number.

Hence, the coordinate of the point in question is (s, s), and

$s = \sin\dfrac{\pi}{4} = \cos\dfrac{\pi}{4}$.   Since $\sin^2 x + \cos^2 x = 1$, we have

$$\sin^2\frac{\pi}{4} + \cos^2\frac{\pi}{4} = s^2 + s^2 = 2\,s^2 = 1.$$

Hence $s^2 = \dfrac{1}{2}$ and $s = \pm\dfrac{1}{\sqrt{2}}$.

Since it is clear that $s > 0$, we have $s = \dfrac{1}{\sqrt{2}}$, and so

$$\sin \frac{\pi}{4} = \frac{1}{\sqrt{2}} \quad . \quad \blacksquare$$

<u>Remark:</u> <u>Memorize this triangle.</u> (It will simplify your life.)

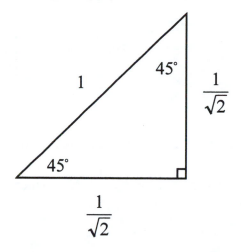

Example 2:   Find $\tan \dfrac{\pi}{6}$.

Solution: Since $\pi/6 = 30°$, the picture now looks like

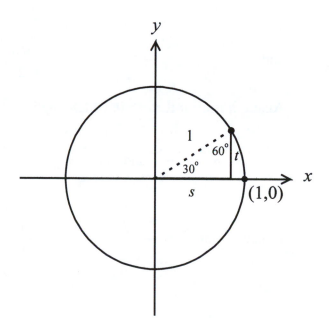

How big is s? And t ? Here's the trick: flip down the triangle along its base, as shown.

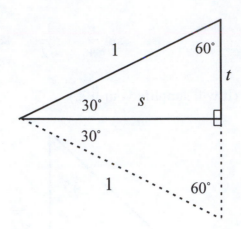

Notice that the result is an equilateral triangle, so the vertical side is also 1.

This means $2t = 1$, or $t = \dfrac{1}{2}$. Now since

$$s^2 + t^2 = 1,$$

$$s^2 + \frac{1}{4} = 1,$$

$$s^2 = \frac{3}{4},$$

and

$$s = \pm\frac{\sqrt{3}}{2}.$$

Again, it is clear that $s > 0$, which implies

$$\cos\frac{\pi}{6} = \frac{\sqrt{3}}{2} \quad \text{and} \quad \sin\frac{\pi}{6} = \frac{1}{2}.$$

Hence $\quad \tan\dfrac{\pi}{6} = \dfrac{\sin\dfrac{\pi}{6}}{\cos\dfrac{\pi}{6}} = \dfrac{\dfrac{1}{2}}{\dfrac{\sqrt{3}}{2}} = \dfrac{1}{\sqrt{3}}.$ ■

<u>Remark:</u> <u>Memorize this triangle also.</u> (This is the last one – we promise!)

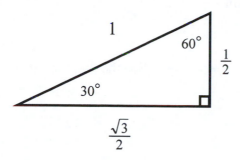

The two triangles we asked you to memorize can be used to determine trigonometric values for many more angles. Consider the following example.

Example 3:   Find $\csc \dfrac{-2\pi}{3}$.

Solution: Since $-\dfrac{2\pi}{3} = -120°$, the triangle in the figure below is exactly the $30°, 60°, 90°$ triangle shown above.

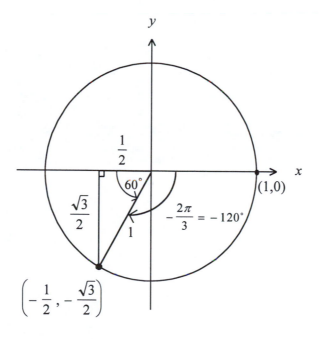

Now since

$$\csc \frac{-2\pi}{3} = \frac{1}{\sin \dfrac{-2\pi}{3}}$$

$$= \frac{1}{-\dfrac{\sqrt{3}}{2}} = \frac{-2}{\sqrt{3}} . \quad \text{(See the diagram above.)} \ \blacksquare$$

Exercises 6.5 Evaluate the following:

1) $\sin \dfrac{3\pi}{4}$ 2) $\sec \dfrac{-2\pi}{3}$ 3) $\csc \dfrac{7\pi}{6}$ 4) $\tan \dfrac{4\pi}{3}$

5) $\cos \dfrac{-3\pi}{4}$ 6) $\sin \dfrac{-3\pi}{4}$ 7) $\cot \dfrac{19\pi}{4}$ 8) $\sec \dfrac{4\pi}{3}$

9) $\csc \dfrac{4\pi}{3}$

Chapter 7

Composition and Decomposition of Functions

7.1 Composition

Consider the function $k(x) = \sqrt{x}$. Then

$$k(1) = \sqrt{1} = 1,$$

$$k(x+2) = \sqrt{x+2},$$

$$k(x^4) = \sqrt{x^4} = x^2,$$

$$k(x^2+x+1) = \sqrt{x^2+x+1},$$

$$k(\sin x) = \sqrt{\sin x},$$

$$k(x+h) = \sqrt{x+h},$$

and

$$k(\text{whatever}) = \sqrt{\text{whatever}}.$$

Get it? Got it? Good!

Now consider the functions $f(x) = \sqrt{x}$ and $g(x) = \sin x$. Then $f(g(x)) = \sqrt{g(x)} = \sqrt{\sin x}$. (Agreed?) Why are we plugging one function into another one? Because it happens all the time out there, in "real life."

Example 1:   Let $f(x) = x^2 + 2x + \pi$ and $g(x) = x^3$. Then what is $f(g(x))$?

Solution: $f(g(x)) = (g(x))^2 + 2g(x) + \pi$

$$= (x^3)^2 + 2x^3 + \pi$$

$$= x^6 + 2x^3 + \pi. \quad \blacksquare$$

Example 2:   If $f(x) = \tan x$ and $g(x) = \sqrt{x+1}$, then what is $f(g(x))$ and $g(f(x))$?

Solution: a) $f(g(x)) = \tan(g(x)) = \tan\sqrt{x+1}$.

b) $g(f(x)) = \sqrt{f(x) + 1} = \sqrt{\tan x + 1}$.

Notice that $f(g(x)) \neq g(f(x))$. ∎

Definition: $f(g(x))$ is called the <u>composition of f with g</u>. Its symbol is $f \circ g$, and its evaluation at a point x is denoted $(f \circ g)(x)$.

Example 3:   Let $f(x) = \dfrac{1}{x}$, and $g(x) = \sin x$. Find $(f \circ g)(x)$.

Solution: $(f \circ g)(x) = f(g(x)) = \dfrac{1}{g(x)} = \dfrac{1}{\sin x} = \csc x$. ∎

Example 4:   Let $f(x) = \dfrac{1}{x+2}$, and $g(x) = x^2 - 1$. Find $(f \circ g)(x)$ and $(g \circ f)(x)$.

Solution: a) $(f \circ g)(x) = f(g(x)) = \dfrac{1}{g(x)+2} = \dfrac{1}{x^2 - 1 + 2} = \dfrac{1}{x^2 + 1}$.

b) $(g \circ f)(x) = (f(x))^2 - 1 = \left(\dfrac{1}{x+2}\right)^2 - 1$.

Also in this example, $f \circ g \neq g \circ f$. ∎

We can define $f \circ g \circ h$ in a similar fashion, and also $f \circ g \circ h \circ k$, and so on, and so on.

Example 5:   Let $f(x) = x^2$, $g(x) = \sin x$, and $h(x) = 2x + 1$. What is the composition $(f \circ g \circ h)(x)$?

Solution: $(f \circ g \circ h)(x) = f\big(g(h(x))\big)$

$$= f\big(g(2x+1)\big) = f\big(\sin(2x+1)\big)$$

$$= \big(\sin(2x+1)\big)^2 = \sin^2(2x+1). \quad \blacksquare$$

Example 6:   Let $f(x) = x^3$, $g(x) = \cos x$, $h(x) = \sqrt{x}$, and $k(x) = x+2$. Find $(f \circ g \circ h \circ k)(x)$.

Solution: $(f \circ g \circ h \circ k)(x) = f\big(g(h(k(x)))\big)$

$$= f\big(g(h(x+2))\big) = f\big(g(\sqrt{x+2})\big)$$

$$= f\big(\cos\sqrt{x+2}\big)$$

$$= \big(\cos\sqrt{x+2}\big)^3 = \cos^3\sqrt{x+2}. \quad \blacksquare$$

Example 7:   Let $f(x) = \cos x$ and $g(t) = t^2$. Find the function $(f \circ g)(t)$.

Solution: Notice here that g is a function of t, not x, but that's no problem. You simply substitute $g(t)$ into $f(x)$ to get

$$(f \circ g)(t) = f\big(g(t)\big) = f(t^2) = \cos(t^2) = \cos t^2. \quad \blacksquare$$

Exercises 7.1

1)   Given the functions

$$f(x) = x^3, \qquad g(x) = \cos x, \qquad s(t) = 2t + 1, \qquad h(x) = \sin x - 4x,$$

find the following composition functions:

a) $(f \circ g)(x)$ b) $(f \circ s)(t)$ c) $(f \circ h)(x)$

d) $(g \circ f)(x)$ e) $(g \circ g)(x)$ f) $(g \circ h)(x)$

2)   Suppose that $f(x) = x^2 - 2x$, $g(x) = \sqrt{x}$, and $h(x) = \tan x$. Find:

a) $(f \circ g \circ h)(x)$ b) $(f \circ h \circ g)(x)$ c) $(g \circ h \circ f)(x)$

d) $(g \circ f \circ h)(x)$ e) $(f \circ f \circ g)(x)$ f) $(g \circ g \circ f)(x)$

7.2 Decomposition

When applying the chain rule to a function, it is necessary to decompose that function, that is, to write it as a composition of simpler functions. In other words, a function that is a composite of two or more other functions will have to be recognized as such. By the way, in the function $f(g(x))$, f is called the outer function, and g is called the inner function. Taking derivatives of functions that are composites requires using the chain rule. It is important to let f be the outermost function (because there may be several ways of decomposing). How can you find the outermost function? Answer: if you were to evaluate the function at some point, the last operation you would do corresponds to the outermost function.

Example 1:   Decompose the function $y(x) = (x^2 + 1)^5$

Solution: If you were to evaluate this function (at $x = 0$ for example), the last operation would be to take the fifth (in a manner of speaking). So $f(x) = x^5$ is the outermost function. Clearly $g(x) = x^2 + 1$ is the inner function and $(f \circ g)(x) = (g(x))^5 = (x^2 + 1)^5$. Hence $f(x) = x^5$, $g(x) = x^2 + 1$ gives the desired decomposition. ∎

Example 2:   Decompose the function $y(x) = \cos^3 x$.

Solution: Since $\cos^3 x$ means $(\cos x)^3$, the last operation would be to take the cube. So $f(x) = x^3$ is the outer function and $g(x) = \cos x$ is the inner function. ∎

Example 3:   Decompose the function $y(x) = \tan \sqrt{x}$.

Solution: Here you would take x, take its root, and then take $\tan$ of the result. So $f(x) = \tan x$ is the outer function, and $g(x) = \sqrt{x}$ is the inner function. ∎

Example 4:   Decompose the function $y(x) = \cos\sqrt{x^2 + 1}$.

Solution: The outermost function is $f(x) = \cos x$, and the inner function is $g(x) = \sqrt{x^2 + 1}$. (By the way, for the next step in using the chain rule, the function $g(x) = \sqrt{x^2 + 1}$ must itself be decomposed, which yields the outer function $\sqrt{x}$ and the inner function $x^2 + 1$.) ∎

Exercises 7.2

Find decompositions for the following functions – i.e., find f and g such that $y = f \circ g$, and f is the outermost function.

1) $y(x) = \tan^2 x$

2) $y(x) = (x^3 - 1)^2$

3) $y(x) = \sin\sqrt{x}$

4) $y(x) = \cos x^5$

5) $y(x) = (\sqrt[5]{x} - 1)^{2/3}$

6) $y(x) = \sin\sqrt{x+1}$

7) $y(x) = \tan^3 2x$

8) $y(x) = \cos(x^3 - 2)^{2/7}$

Chapter 8

Equations of Degree 1 and 2 (mostly)

8.1 Equations of Degree 1

You're asking: why are we doing this? This is the easiest thing in algebra. Well, it is, but it can get confusing when too many different variables are floating about. But as long as you keep things straight, it really is easy.

Example 1: Solve for x: $\frac{2}{3}x - 4 = 0$.

Solution: Get rid of the -4 by adding 4 to both sides. This gives

$$\frac{2}{3}x = 4.$$

Now, get rid of the $\frac{2}{3}$ by multiplying both sides by $\frac{3}{2}$, and get

$$x = \frac{3}{2} \cdot 4 = 6. \text{ Easy! } \blacksquare$$

Example 2: Solve for y: $\pi y - 4 + 2\pi = 0$.

Solution: While this looks a little more complicated, y appears in only one place, so you can use a similar method (we call it "peeling the onion"). Keep stripping off terms, or factors, until the desired variable is exposed. In this case, you first "peel away" the $-4 + 2\pi$ by subtracting it from both sides to get

$$\pi y = 4 - 2\pi.$$

The next layer to peel away is that π on the left side of the equation. Divide by π to get

$$y = \frac{4 - 2\pi}{\pi}. \quad \blacksquare$$

Example 3:   Solve for x: $2y^2 x + (yw^2 - 2)^2 = 0$.

Solution: While this looks a little more complicated, x appears in only one place, so you can start peeling. In this case, first "peel away" the $(yw^2 - 2)^2$ by subtracting it from both sides to get

$$2y^2 x = -(yw^2 - 2)^2.$$

Next peel off the $2y^2$ by division, resulting in

$$x = -\frac{(yw^2 - 2)^2}{2y^2}. \quad \blacksquare$$

Remark: The peeling away of the layers can be used when the variable occurs in only one place. It can be used not only in equations of degree 1 but also in many others.

Example 4:   Solve for x: $(\sqrt{x} + 2)^3 - 64 = 0$.

Solution: This is not of degree 1, but it is of "onion type" because x is in one place only. To get at it, remove the outermost layer by adding 64 to both sides:

$$(\sqrt{x} + 2)^3 = 64.$$

Next, remove the cube by taking the cube root

$$\sqrt{x} + 2 = 64^{\frac{1}{3}} = 4,$$

remove the 2,

$$\sqrt{x} = 4 - 2 = 2,$$

and square both sides

$$x = 4.$$

You must be very careful to check your solution, or solutions, as the case might be. (Do it now!) $\blacksquare$

Here are some other examples of equations you might meet.

Example 5:   Solve for x: $2x + 5y = 3x + y + 1$.

Solution: Here x occurs more than once, but that can be fixed by subtracting $3x$:

$$-x + 5y = y + 1.$$

Now, subtract $5y$ and multiply by -1 to get

$$x = 4y - 1. \blacksquare$$

In calculus, you will deal with things called derivatives. One way of denoting the derivative is the symbol $\dfrac{dy}{dx}$. Think of it as just the name of a variable, like x, or y, or w, or Δx. Another symbol for the derivative is y'.

Example 6:   Solve for $\dfrac{dy}{dx}$: $5xy + 4\dfrac{dy}{dx} = 3x^2 - 2xy^2\dfrac{dy}{dx}$.

Solution: Isolate the variable you wish to solve for, which in this case is the derivative $\dfrac{dy}{dx}$. Bring all the terms with $\dfrac{dy}{dx}$ to the left, and all others to the right, by adding and subtracting as needed. We get

$$4\dfrac{dy}{dx} + 2xy^2\dfrac{dy}{dx} = 3x^2 - 5xy,$$

which we combine by taking out the common factor $\dfrac{dy}{dx}$, to get

$$\dfrac{dy}{dx}(4 + 2xy^2) = 3x^2 - 5xy.$$

Division by $(4 + 2xy^2)$ gives the result

$$\dfrac{dy}{dx} = \dfrac{3x^2 - 5xy}{4 + 2xy^2}. \blacksquare$$

Example 7: Solve for y': $x^2 y' + 2xy + 2xyy' + y^2 + (\cos y)y' = 0$.

Solution: Again, move all the terms with y' to the left, and all others to the right:

$$x^2 y' + 2xyy' + (\cos y)y' = -2xy - y^2.$$

Factor out the common factor y':

$$\left(x^2 + 2xy + \cos y\right)y' = -2xy - y^2.$$

Divide by $\left(x^2 + 2xy + \cos y\right)$, to get

$$y' = \frac{-2xy - y^2}{\left(x^2 + 2xy + \cos y\right)}. \quad \blacksquare$$

If the equation is not of degree 1, then you must take great care to make sure the calculated solution is in fact valid. That is, does it satisfy the original equation? Some "solutions" do not satisfy the given equation, and hence are NOT solutions, as the following example illustrates.

Example 8:   Solve $\dfrac{1}{x-5} + \dfrac{1}{x+5} = \dfrac{10}{x^2 - 25}$ for x.

Solution: Multiplying by $x^2 - 25$ will give you

$$x + 5 + x - 5 = 10,$$

or $2x = 10$ and $x = 5$.

But $x = 5$ cannot be used in the original equation without dividing by zero. Hence there is no solution to this problem. $\blacksquare$

Exercises 8.1

1) Solve for x: $\dfrac{5}{12}x - 35 = 0$.

2) Solve for x: $\dfrac{2}{3}x - 3 = \dfrac{1}{5}x + 2$.

3) Solve for y: $2zy + 3z + \pi = 0$.

4) Solve for x: $\pi - 4z^2y^2 + (2zy^2 + 2y)x = 0$.

5) Solve for $\dfrac{dy}{dx}$: $2xy + x^2\dfrac{dy}{dx} + 3x^2y^3 + 3x^3y^2\dfrac{dy}{dx} = 0$.

6) Solve for $\dfrac{dy}{dx}$: $xy + x\dfrac{dy}{dx} + 2xy^2 + 2x^2y\dfrac{dy}{dx} = 3x - 2y\dfrac{dy}{dx}$.

7) Solve for y': $\sin x + y'\cos x = 2xy' - 1$.

8.2 Equations of Degree 2

You won't need this tool for implicit differentiation, but it is a basic addition for your toolbox. Don't leave home without it! As already seen in Chapter 2, the solutions to the quadratic equation

$$a x^2 + b x + c \;=\; 0, \qquad \text{with } a \neq 0,$$

are

$$x = \frac{-b \pm \sqrt{b^2 - 4ac}}{2a}.$$

This is one of the most useful formulas you will meet. **MEMORIZE IT NOW!!** If you don't, you will stray from the straight and narrow, and not even Elvis will be able to save you.

Example 1:   Solve $x^2 - 3x + 2 = 0$ for x.

Solution: Here $a = 1$, $b = -3$, and $c = 2$, and $b^2 - 4ac = 1$. So, according to the quadratic formula

$$x = \frac{3 \pm \sqrt{9 - 8}}{2} = \frac{3 \pm 1}{2}$$

$$= 2 \text{ or } 1.$$

So there are two solutions: $x = 1$ and $x = 2$. Can you see how useful this formula is? ∎

Example 2:   Solve $s^2 + 4s + 4 = 0$ for s.

Solution: Don't let the s fool you, this is just a quadratic equation in s, and s takes the place of x in the preceding discussion. Here $a = 1$, $b = 4$, and $c = 4$. So,

$$s = \frac{-4 \pm \sqrt{16 - 16}}{2} = \frac{-4 \pm 0}{2} = -2.$$

In this case there is only one root. When this happens, the root is called repeated, or a root of order 2. (When speaking of equations, root is another word for solution.) It means that the original quadratic could be factored as $s^2 + 4s + 4 = (s + 2)^2$, and yes, you guessed it. The case of repeated roots correspond to quadratic polynomials that are in fact perfect squares. ■

Sometimes a quadratic equation can't be solved in terms of real numbers. Consider the following.

Example 3:   Solve for y: $y^2 + 2y + 2 = 0$.

Solution: Here $a = 1$, $b = 2$, and $c = 2$, so that

$$y = \frac{-2 \pm \sqrt{4-8}}{2} = \frac{-2 \pm \sqrt{-4}}{2} = \frac{-2 \pm 2\sqrt{-1}}{2} = -1 \pm \sqrt{-1}.$$

The term $\sqrt{-1}$ is not defined as a real number, because no real number when squared gives a negative number. So there are no real solutions. ■

Remark: If $b^2 - 4ac < 0$, as in the last example, there are no real solutions. But wishful thinking is a great motivator, and in this spirit, mathematicians several centuries ago agreed to consider $\sqrt{-1}$ an "imaginary number." We represent $\sqrt{-1}$ by the symbol i (for imaginary). In this case the roots of the last equation could be written as $-1 \pm i$.

Any number of the form $a + ib$ is called a complex number, and all four basic operations (addition, subtraction, multiplication, and division) carry over from real numbers, provided you remember $i^2 = -1$. (In the 1800s, complex numbers were removed from the realm of hocus-pocus and put onto a logically sound foundation – which means that today we can all sleep much easier.) Now check the numbers $-1 \pm i$ to see if they are in fact solutions to the problem in Example 3.

Example 4: Verify that $y = -1 \pm i$ are solutions to $y^2 + 2y + 2 = 0$.

Solution: We will check one part at a time.

a) If $y = -1 + i$, then

$$y^2 = (-1+i)^2 = (-1)^2 - 2i + i^2$$
$$= 1 - 2i - 1 = -2i$$
$$2y = 2(-1+i) = -2 + 2i.$$

Hence $y^2 + 2y + 2 = -2i - 2 + 2i + 2 = 0$. So it works!

b) If $y = -1 - i$, then

$$y^2 = (-1-i)^2 = (-1)^2 + 2i + (-i)^2$$
$$= 1 + 2i - 1 = +2i.$$
$$2y = 2(-1-i) = -2 - 2i.$$

Hence $y^2 + 2y + 2 = 2i - 2 - 2i + 2 = 0$. ■

Remark: These examples demonstrate the three possible outcomes when solving quadratic equations. The number $b^2 - 4ac$ is called the <u>discriminant</u>, since it allows you to "discriminate" among the following three cases that indicate what types of roots you will have.

1) $b^2 - 4ac > 0$: In this case, the quadratic equation has two real and distinct roots – that is, two roots that are different numbers. This was the situation in Example 1.

2) $b^2 - 4ac = 0$: Here, the square root in the solution disappears, leaving one real repeated root. This happened in Example 2.

3) $b^2 - 4ac < 0$: In this last case, the roots are complex numbers, as was the case in Example 3.

It is possible that in your travels you will meet up with a quadratic equation that is trying to disguise itself. Take the following example.

Example 5:   Solve for x: $zx + 2yx^2 + yx + z^2 - y^2 = 0$.

Solution: On first appearance this seems formidable, but by rewriting the equation as

$$(2y)x^2 + (z+y)x + (z^2 - y^2) = 0$$

you can that see this is a quadratic equation in x with $a = 2y$, $b = z+y$, and $c = z^2 - y^2$. Hence plugging these values into the quadratic formula gives

$$x = \frac{-(z+y) \pm \sqrt{(z+y)^2 - 8y(z^2 - y^2)}}{4y}.\ \blacksquare$$

Sometimes a higher order equation can be reduced to a quadratic by a simple substitution. Consider the following case.

Example 6:   Solve for x: $x^4 - 5x^2 + 4 = 0$.

Solution: If we use the substitution $y = x^2$, then this equation becomes

$$y^2 - 5y + 4 = 0,$$

which is quadratic and has roots $y = 1$ and $y = 4$. Hence the solutions of the original equation are $x^2 = 1$ and $x^2 = 4$, giving $x = \pm 1$ and $x = \pm 2$. $\blacksquare$

You have to be very careful however: since you are taking square roots, some solutions will not be real valued. Consider the following example:

Example 7:   Solve for x: $x^4 - 5x^2 - 36 = 0$.

Solution: Again, letting $y = x^2$, you will obtain the quadratic equation

$$y^2 - 5y - 36 = 0,$$

which has roots $y = -4$ and $y = 9$. Hence the solutions of the original

equation are $x^2 = -4$ and $x^2 = 9$, giving $x = \pm 2i$ and $x = \pm 3$.

However, if the original problem needs x to be real, you must only consider

the solutions $x = \pm 3$. ∎

If an equation in x has three terms, one constant, one like x^m, and the last like x^{2m}, then the substitution $y = x^m$ leads to a quadratic equation in y. Consider the following example.

Example 8:   Solve $x^6 + 6x^3 - 16 = 0$ for real x.

Solution: Use the substitution $y = x^3$ to obtain

$$y^2 + 6y - 16 = 0,$$

which has solutions

$$y = -8, \quad \text{and} \quad y = 2,$$

so $$x^3 = -8, \quad \text{and} \quad x^3 = 2,$$

or $$x = -2, \quad \text{and} \quad x = \sqrt[3]{2} = 2^{\frac{1}{3}}. \quad ∎$$

For more complicated equations with coefficients that are also variables, the method is the same, but you should always verify (i.e., check) the solutions.

Exercises 8.2 Find the solutions, both real and complex, for exercises 1 to 9.

1) $x^2 + 5x + 4 = 0$ 2) $g^2 - 4g + 5 = 0$ 3) $x^2 + 6x + 9 = 0$

4) $y^2 + 7y + 4 = 0$ 5) $y^2 - 8y + 16 = 0$ 6) $z^4 - 8z^2 + 16 = 0$

7) $y^2 - 64 = 0$ 8) $x^4 - 64 = 0$ 9) $s^4 - 9s^2 + 20 = 0$

10)   Find real solutions to $x^6 - 26x^3 - 27 = 0$.

11)   Solve $\frac{1}{2}x^2 + (y + z)x + z^2 - y^2 = 0$ for x.

12)   Solve $\frac{1}{2}x^2 + (y + z)x + z^2 - y^2 = 0$ for z.

13)   Solve for x: $x + 3\sqrt{x} - 10 = 0$. (Hint: Let $w = \sqrt{x}$. Check your solutions!)

14)   Solve for x in $[0, 2\pi]$: $\sin^2 x - \sin x = 0$.

15)   Solve for x: $2\cos^2 x + \cos x - 1 = 0$, for x in $[-\pi, \pi]$.

Chapter 9

Word Problems, Algebraic and Trigonometric

9.1 Algebraic Word Problems

The methods of finding maxima and minima are some of the most useful mathematical tools in calculus. After all, in many situations, there is a desire to minimize cost, maximize gain, maximize volume for a given amount of material, or minimize the material necessary to enclose a given volume. It is one of the main goals in engineering, science, business, and is often called optimization (how to do something in the best possible way).

The so-called max-min problems are phrased in a way that is intended to reflect a real-life situation as it may occur to an engineer, for example. First, it is necessary to interpret the situation, to see what quantities are involved, and what questions need to be asked and answered. You won't be told: optimize this specific function, but rather you'll first have to figure out, on your own, what function needs to be optimized. The point is that you need to first understand the problem as given, and to figure out what the variables are and how they are related to each other as functions. "A picture is worth a thousand words," and nine times out of ten, the situation is made easier to understand with a well-drawn diagram. You should make a habit of interpreting word problems through visualization. What does the situation look like in your mind's eye? Now draw it.

Example 1:   A rectangular field is to be fenced off next to a straight river, with fencing on three sides, with the river edge making the fourth side. Exactly 100 feet of fencing is to be used. Express the area of the field as a function of its width.

Solution: First draw a diagram and label the edges.

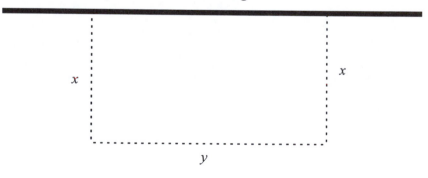

Let x = the width of the field.

Let y = the length of the field.

Note that $x + y + x = 2x + y = 100$, so $y = 100 - 2x$.

Hence the area $A = xy = x(100 - 2x) = 100x - 2x^2$. ∎

Example 2:   A swimming pool is in the shape of a square with a semicircle at each of two opposite edges. Express the perimeter and area of the pool as a function of the diameter of the semicircles.

Solution:

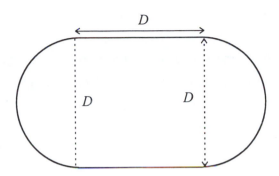

a)   The circumference of a circle is $2\pi r$, so the sum of the arc lengths of the two semicircles is $2\pi r$, which equals πD. The square contributes $2D$, so the perimeter of the pool is $\pi D + 2D$.

b)   The area of a circle is πr^2, so the sum of the area of the two semicircles is $\pi r^2 = \dfrac{\pi D^2}{4}$, making the area of the pool equal to $\dfrac{\pi D^2}{4} + D^2$.   ∎

Example 3:   A cylindrical tin can has height h and radius r. Its volume is 32 fluid ounces. Express h as a function of r, and vice versa.

Solution: Again, first draw a picture and label the variables.

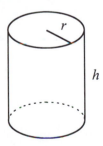

$$V = (\text{ area of end })(\text{height}) = (\pi r^2)h = 32.$$

$$\therefore\ h = \frac{32}{\pi r^2}.$$

Also $\pi r^2 = \dfrac{32}{h}$, so $r^2 = \dfrac{32}{\pi h}$, and hence

$$r = \sqrt{\dfrac{32}{\pi h}} \, . \qquad \blacksquare$$

Example 4:   A 10-inch wire is cut into two pieces. One of the pieces, of length x, is bent into a square. The other piece is bent into a circle. What is the total area of the two shapes as a function of x?

Solution: Since one piece is of length x inches, the other piece is of length $10 - x$ inches.

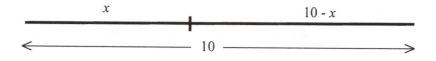

The piece of length x is bent into a square of side $x/4$, and therefore the area of the square is $\dfrac{x^2}{16}$ in². The second piece of wire is of length $10 - x$, which forms the circumference of the circle. Hence we have $10 - x = 2\pi r$, and $r = \dfrac{10 - x}{2\pi}$. So the area of the circle is $\pi r^2 = \pi \left(\dfrac{10 - x}{2\pi} \right)^2 = \dfrac{(10 - x)^2}{4\pi}$.

The sum of the areas is $A = \dfrac{x^2}{16} + \dfrac{(10 - x)^2}{4\pi}$. $\blacksquare$

Recall that when we say that the variable y is proportional to the variable x, we mean that $y = kx$ for all x, y and for some fixed number k, which is called the proportionality constant. Instead of saying "is proportional to" we can also say "varies as." We say that the variable y is inversely proportional to the variable x if $y = k/x$ for all x and y, and for some constant k. When y is said to follow the inverse square law with respect to x, we mean that $y = k/x^2$ for some constant k. There are many situations in real life where quantities obey these relationships. Here are two examples.

Example 5:   The cost C of building a highway through a certain section of the country is proportional to its length L. A 2.5-mile section costs \$1 million. Express the cost as a function of the length, and compute the cost of building 13.2 miles.

Solution: The first sentence tells us that $C = kL$, for some fixed k. The constant k can be computed from the information given in the second sentence, which can be rephrased symbolically as $1,000,000 = (k)(2.5)$.

$\therefore k = 400,000$, and $C = 400,000\,L$.

Hence, the cost C of building 13.2 miles is

$C = (400,000)(13.2) = 5,280,000$, and so the cost is \$5,280,000. ∎

Remarks: a) The value of k in the example above depends on the units chosen for cost and length. If you chose dollars and feet, then k would be $\dfrac{C}{L} = \dfrac{1,000,000}{(2.5)(5280)} \cong 75.8$. It is important to be consistent in using units. The final answer however will be the same, no matter which units you chose. (Show this!)

b) Once you have chosen the units you wish to use and have made sure that all quantities are consistently expressed in those units only, we can solve the problem without explicitly writing down the units at each step of the calculation. All calculated quantities will automatically come out in the chosen units.

Example 6:   The gravitational force between two point-masses satisfies the inverse square law with respect to the distance between them. Suppose the gravitational force acting on you at sea level is 150 pounds, and that you and the earth can be considered point-masses with the mass concentrated at the centers. If the radius of the earth is approximately 4000 miles, and Mount Everest is 29,028 feet high, estimate the gravitational force acting on you on top of Mount Everest.

Solution: First of all, 29,028 feet $\cong$ 5.5 miles. We know the force F satisfies the inverse square law with respect to distance, so

$$F = \frac{k}{r^2},$$

where k is some constant, depending on you and the earth, and

$r =$ the distance between you and the center of the earth.

At sea level, you have

$$150 = \frac{k}{4000^2},$$

and so $k = 2,400,000,000 = 2.4 \times 10^9$, and hence

$$F = \frac{2.4 \times 10^9}{r^2}.$$

On top of Mount Everest, $r = 4000 + 5.5 = 4005.5$, and so

$$F = \frac{2.4 \times 10^9}{4005.5^2} = \frac{2.4 \times 10^9}{4.0055 \times 10^6} \cong 0.1496 \times 10^3 = 149.6 \text{ lb.}$$

So the force of gravity pulls you almost half a pound less than it would at sea level, for example in Boston. ■

Exercises 9.1

1) A rectangular field of area 20,000 sq. ft. is to be fenced off next to a river, with fencing on three sides and the river making the fourth side. Express the length of fencing necessary as a function of the width of the field.

2) The shape of a window is given by two squares, one on top of the other, with a semicircle on top of that. Find the perimeter and area of the window as a function of the width of the window.

3) A cylindrical can is to be made up from sheet steel so that its surface area is 100 sq. in. Express the height as a function of the radius. Hint: Imagine removing the top and bottom with a can opener, and splitting the rest down the side and unrolling it flat.

4) An open cardboard box is to be constructed from a rectangular 8" by 10" sheet by cutting identical squares of side x out of each corner, and folding up the resulting edges. Determine the volume of the box as a function of x.

5) An open cardboard box is to be constructed from a rectangular 1.5 m by 2 m sheet by cutting identical squares of side x out of each corner, and folding up the resulting edges. Determine the volume of the box as a function of x. Also, what is the volume of the box as a function of the exterior surface area?

6) The volume of a sphere is $V = \dfrac{4}{3}\pi r^3$, and its surface area is $S = 4\pi r^2$. Express V as a function of S and vice versa.

7) Suppose the gravitational force acting on you at sea level is 140 lb., and that you and the earth can be considered point-masses with the mass concentrated at the centers. If the radius of the earth is approximated at 4000 miles, and the Matterhorn is 14,688 feet high, estimate the gravitational force acting on you on top of the mountain.

8) The distance between the earth and the moon varies from a maximum of approximately 253,000 miles to a minimum of around 221,000 miles. Find the percentage increase in the gravitational force acting between the two bodies when going from farthest separation to closest separation.

9) A 2 ft. wire is cut into two pieces. One of the pieces, of length x, is bent into a circle. The other piece is bent into a rectangle whose length is twice the size of its width. What is the total area of the two shapes as a function of x?

9.2 <u>Right Triangle Trigonometry</u>

In Chapter 6, we introduced the sine and cosine functions of any number θ as the coordinates of a point on the unit circle. For numbers θ between 0 and $\frac{\pi}{2}$ there is another way of defining them as well as the other basic trigonometric functions.

To illustrate, consider any number θ with $0 < \theta < \frac{\pi}{2}$. Notice that θ radians would be an acute angle, so you may draw a right triangle (short for right-angled triangle) with an angle of θ radians, as for example $\triangle ABC$.

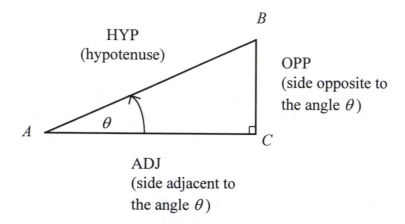

Now place the unit circle onto $\triangle ABC$, with its center at A and the positive x-axis along the line segment AC.

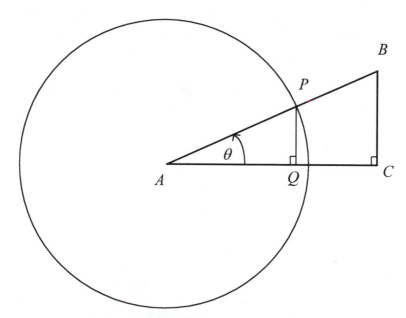

Let the intersection of the circle and the line AB be called P. From P drop the perpendicular to AC, to a point Q. Since P is on the unit circle and the terminal line of the angle θ, we have $P = (\cos\theta, \sin\theta)$

from the definition presented in Chapter 6. Hence, $PQ = \sin\theta$, while $AQ = \cos\theta$ and $AP = 1$. So here we have two triangles,

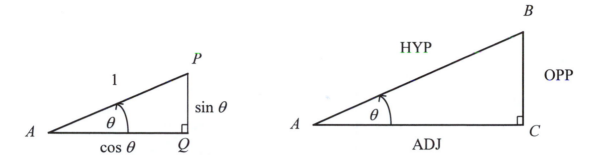

that are in fact similar.

Since they are similar, their sides are in proportion, so

$$\frac{\sin\theta}{1} = \frac{OPP}{HYP},$$

and hence

$$\sin\theta = \frac{OPP}{HYP}.$$

Similarly

$$\frac{\cos\theta}{1} = \frac{ADJ}{HYP},$$

and so

$$\cos\theta = \frac{ADJ}{HYP}$$

Continuing on,

$$\tan\theta \;=\; \frac{\sin\theta}{\cos\theta} \;=\; \frac{OPP}{ADJ}, \quad \cot\theta \;=\; \frac{ADJ}{OPP},$$

$$\sec\theta \;=\; \frac{HYP}{ADJ}, \;\; \text{and} \;\; \csc\theta \;=\; \frac{HYP}{OPP}.$$

Some books use these relationships as the definition of the six basic trigonometric functions.

If you know certain sides and angles of a right triangle, the trigonometric functions can be used to find the rest, which is called <u>solving the triangle</u>. Since $\sin\theta = \dfrac{OPP}{HYP}$ if θ is in $[0, \dfrac{\pi}{2}]$, and if you know the size of θ, you can calculate (for example with a calculator or a table) the value of $\sin\theta$ and hence the ratio $\dfrac{OPP}{HYP}$. You can also get the value of the other five trigonometric functions. So, if you know θ and one of the sides, you can get the other sides also. Of course, you will make full use of the <u>Pythagorean theorem</u>: in a right triangle, the square of the hypotenuse is equal to the sum of the squares of the other two sides. This can be expressed as $(HYP)^2 = (OPP)^2 + (ADJ)^2$.

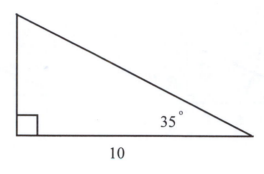

Example 1:   Solve:

Solution: We know that ADJ = side adjacent to the $35°$ angle = 10. To get OPP, the side opposite the angle, consider that in this case $\dfrac{OPP}{ADJ} = \tan 35°$, and so

$OPP = ADJ \cdot \tan 35° \cong (10) \cdot (.7002) = 7.002$. (Don't forget, when using your calculator, to make sure it is set on degrees when using degrees, not radians.)

Given the opposite side, OPP, you can get the hypotenuse, HYP, from the Pythagorean theorem, or by using $\dfrac{HYP}{ADJ} = \sec 35°$, to obtain

$HYP = ADJ \cdot \sec 35° \cong (10) \cdot (1.2208) = 12.208$. Of course, the last angle is $55°$. (Right???) ∎

Example 2:   A rocket blasts off vertically, its path followed by a camera on the ground 2 miles away. Find a relation between the height of the rocket and the angle of elevation of the camera.

Solution: First draw a picture and label the physical quantities. Let the variable h denote the height, or altitude, of the rocket in miles, and let the angle of elevation of the camera be θ.

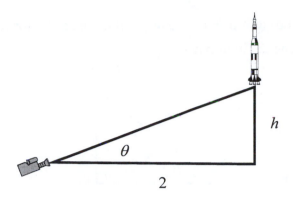

Notice that OPP = h and ADJ = 2, so that $\tan\theta = \dfrac{h}{2}$ and $h = 2\tan\theta$. ∎

Example 3:   Given the triangle below, find α, β, and x.

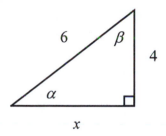

Solution: Finding x is easy: $x = \sqrt{36-16} = \sqrt{20} = 2\sqrt{5}$.

To find α, notice that $\sin\alpha = \dfrac{4}{6} = \dfrac{2}{3} \cong .667$.

This problem is different from what you're used to. Typically, you are given an angle and then asked for the trig value. Here you are given the trig value and asked for the angle. You will learn about this further when you study inverse trig functions. For now, however, you can solve the problem with your calculator. By entering the number .667 and then pressing the $\sin^{-1}$ key, you get $\alpha \cong 42°$, which means that $\beta \cong 48°$. ∎

Exercises 9.2 Consider the right triangle below, and use a calculator to find the angles (to the nearest degree) and the sides (to one decimal place).

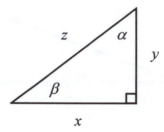

1) If $x = 3$ and $y = 5$, what is z and the angles α and β?

2) If $\alpha = 40°$ and $x = 10$, what are the remaining variables?

3) If $\alpha = 32°$ and $y = 8$, solve the triangle.

4) If $\beta = 54°$ and $x = 5$, find the rest.

5) Solve in terms of h:

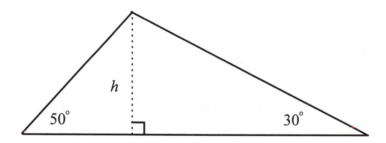

6)   A fire ladder is mounted on its truck at a height of 4 feet. The ladder is 48 feet long, and it can rise at most $70°$ above the horizontal. If the first floor is street level, and each floor is 9 ft. high, assess your chances if you're standing on the sixth floor, looking out the window, in a hurry.

7)   You can walk at 4 mph and row a boat at 2 mph. You are at point A, and need to cross a 3-mile wide river and get to point B, by rowing from A to the shore as shown, and then walking along the shore to B. Point B is 10 miles away from the point directly across the river from A. (Assume that the speed of the current is negligible.)

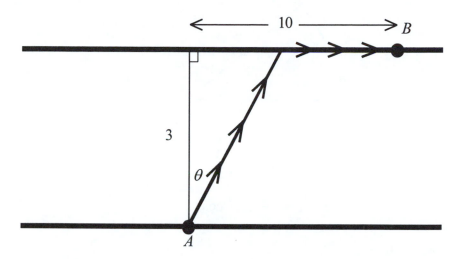

Express the time it will take you as a function of the angle θ. (Recall that $D = R \cdot T$, so $T = \dfrac{D}{R}$ and $R = \dfrac{D}{T}$.)

9.3 The Law of Sines and the Law of Cosines

Consider <u>any triangle</u> whose angles are A, B, and C, and whose sides opposite those angles are a, b, and c, respectively.

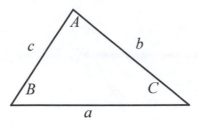

Law of Sines: $$\frac{\sin A}{a} = \frac{\sin B}{b} = \frac{\sin C}{c}.$$

Law of Cosines: $$c^2 = a^2 + b^2 - 2ab\cos C,$$

which also means that

$$a^2 = b^2 + c^2 - 2bc\cos A$$

and

$$b^2 = a^2 + c^2 - 2ac\cos B.$$

In the law of cosines, what happens if the angle is $90°$? Does the equation look familiar? These laws are useful for solving acute or scalene triangles.

Example 1:   Solve:

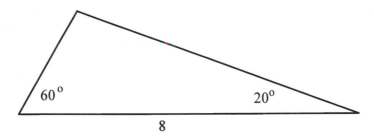

Solution: The law of sines tells all.

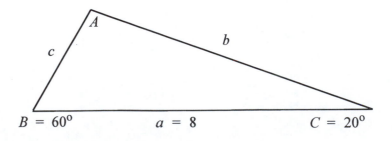

Notice that $A = 100°$, so

$$\frac{\sin 100°}{8} = \frac{\sin 60°}{b} = \frac{\sin 20°}{c}.$$

You can solve for b and c as

$$b = \frac{8 \sin 60°}{\sin 100°} \quad \text{and} \quad c = \frac{8 \sin 20°}{\sin 100°}.$$

Using a calculator or table, you can get decimal equivalents.

(Don't forget to set your calculator to degrees rather than radians.) ∎

Example 2:   Solve

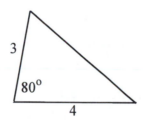

Solution: The law of cosines looks more promising here. First, label the triangle:

$c = 3$, b, C, A, $B = 80°$, $a = 4$

So, $b^2 = a^2 + c^2 - 2ac\cos B$

$\qquad = 16 + 9 - 24\cos 80°$

$\qquad \cong 20.8$

$\therefore b \cong 4.56.$

Now you can find the angle C from the law of sines:

$$\frac{\sin B}{b} = \frac{\sin C}{c}, \quad \text{so} \quad \frac{\sin 80°}{4.56} = \frac{\sin C}{3} \quad \text{and}$$

$$\sin C = 3 \cdot \frac{\sin 80°}{4.56} \cong .648.$$

$\therefore$ C = angle, between 0 and 90, whose sine is .648. Using a calculator we find out that $C \cong 40°$.

$$\therefore A \cong 60°. \quad \blacksquare$$

Exercises 9.3 Consider the triangle below.

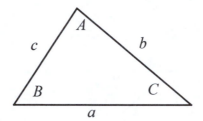

1) If $A = 100°$, $B = 50°$, $b = 12$, find the rest.

2) If $a = 6$, $b = 5$, $C = 60°$, solve the triangle.

3) If $a = 4$, $b = 5$, $c = 6$, solve the triangle.

4) Let $C = 20°$, $c = 2$, $b = 5$. Find two triangles with these measures. Draw the triangles.

5) What results when the angle mentioned in the law of cosines is $90°$?

Chapter 10

Trigonometric Identities

The basic idea is that antidifferentiating $\tan^2 x$ is really tough until you realize that $\tan^2 x = \sec^2 x - 1$, and then there's nothing to it, because you know how to integrate $\sec^2 x$ and 1. That's the point: change what you have into something that you can handle. Chapter 6 focused on the basic trigonometric identities. Here's a list:

Basic Identities

$$\tan x = \frac{\sin x}{\cos x}, \qquad \cot x = \frac{\cos x}{\sin x}$$

$$\sec x = \frac{1}{\cos x}, \qquad \csc x = \frac{1}{\sin x}$$

Circular Identities

$$\boxed{\sin^2 x + \cos^2 x = 1}$$ which implies

$$\sin^2 x = 1 - \cos^2 x, \quad \text{so} \quad \sin x = \pm\sqrt{1 - \cos^2 x}$$

$$\cos^2 x = 1 - \sin^2 x, \quad \text{so} \quad \cos x = \pm\sqrt{1 - \sin^2 x}$$

Notice that for $\sin x = \pm\sqrt{1 - \cos^2 x}$, you can't say whether it is "+" or "−", unless you have some additional information – for example, the quadrant that x is in. If you divide both sides of the equation $\sin^2 x + \cos^2 x = 1$ by $\cos^2 x$, you get

$$\boxed{\tan^2 x + 1 = \sec^2 x},$$

which gives

$$\tan^2 x = \sec^2 x - 1.$$

Taking square roots gives the following equations:

$$\tan x = \pm\sqrt{\sec^2 x - 1}$$

and

$$\sec x = \pm\sqrt{\tan^2 x + 1}.$$

Again the sign may be determined if you have extra information.

Lastly, dividing both sides of the equation $\sin^2 x + \cos^2 x = 1$ by $\sin^2 x$ gives

$$\boxed{1 + \cot^2 x = \csc^2 x}\,,$$

which gives
$$\cot^2 x = \csc^2 x - 1.$$

Taking square roots now gives

$$\cot x = \pm \sqrt{\csc^2 x - 1}\,,$$

and
$$\csc x = \pm \sqrt{1 + \cot^2 x}\;.$$

So much for the old news and modified old news. Here's some new news:

Addition (or Sum) Formulas (MUST BE MEMORIZED):

$$\boxed{\sin(A + B) = \sin A \cos B + \cos A \sin B}$$

$$\boxed{\cos(A + B) = \cos A \cos B - \sin A \sin B}$$

These formulas have several often-used consequences:

a) $\sin 2x = \sin(x + x)$

$\qquad = \sin x \cos x + \cos x \sin x$

$$\boxed{\sin 2x = 2 \sin x \cos x}$$

b) $\cos 2x = \cos(x + x)$

$\qquad = \cos x \cos x - \sin x \sin x$

$$\boxed{\cos 2x = \cos^2 x - \sin^2 x}$$

c)   Since $\cos 2x = \cos^2 x - \sin^2 x$, we have

$\quad \cos 2x = (1 - \sin^2 x) - \sin^2 x = 1 - 2\sin^2 x$, which gives

$$\boxed{\sin^2 x = \frac{1 - \cos 2x}{2}} \quad \text{and so} \quad \boxed{\sin x = \pm \sqrt{\frac{1 - \cos 2x}{2}}}.$$

Also, since $\cos 2x = \cos^2 x - \sin^2 x$, we have $\cos 2x = \cos^2 x - (1 - \cos^2 x) = 2\cos^2 x - 1$, which gives

$$\boxed{\cos^2 x = \frac{1 + \cos 2x}{2}} \quad \text{and so} \quad \boxed{\cos x = \pm \sqrt{\frac{1 + \cos 2x}{2}}}.$$

The equations involving $\sin^2 x$ and $\cos^2 x$ will come in handy when doing antidifferentiation. The other two equations with the square roots provide a nice way to determine sine and cosine of an angle, knowing the cosine of twice the angle. Finally, by substituting $-B$ for B into the addition formulas, you get the difference formulas.

d)   Since $\sin(A - B) = \sin(A + (-B))$,

$$\sin(A - B) = (\sin A)(\cos(-B)) + (\cos A)(\sin(-B))$$

$$= (\sin A)(\cos B) + (\cos A)(-\sin B)$$

or $$\boxed{\sin(A - B) = \sin A \cos B - \cos A \sin B}$$

e)   Since $\cos(A - B) = \cos(A + (-B))$,

$$\cos(A - B) = (\cos A)(\cos(-B)) - (\sin A)(\sin(-B))$$

$$= (\cos A)(\cos B) - (\sin A)(-\sin B)$$

or $$\boxed{\cos(A - B) = \cos A \cos B + \sin A \sin B}$$

<u>Remark:</u>   You can either memorize the results of parts (a) through (e), or recall that it all comes from the addition formulas, while remembering that for parts (d) and (e)

$$\cos(-x) = \cos x,$$

and $$\sin(-x) = -\sin x.$$

Example 1:   Without using a calculator, determine $\sin 75°$.

Solution: Since you can write $75° = 30° + 45°$,

$$\sin 75° = \sin(30° + 45°) = \sin 30° \cos 45° + \sin 45° \cos 30°$$

using the addition formula. Since $\sin 30° = \dfrac{1}{2}$, $\cos 30° = \dfrac{\sqrt{3}}{2}$, and $\sin 45° = \cos 45° = \dfrac{\sqrt{2}}{2}$ we have

$$\sin 75° = \frac{1}{2} \cdot \frac{\sqrt{2}}{2} + \frac{\sqrt{2}}{2} \cdot \frac{\sqrt{3}}{2} = \frac{\sqrt{2}}{4}(1 + \sqrt{3}). \quad \blacksquare$$

Example 2:   Without using a calculator, determine $\cos 15°$.

Solution: Since $15 = 45 - 30$, you have

$$\cos 15° = \cos(45° - 30°) = \cos 45° \cdot \cos 30° + \sin 45° \cdot \sin 30°$$

or $\cos 15° = \dfrac{\sqrt{2}}{2} \cdot \dfrac{\sqrt{3}}{2} + \dfrac{\sqrt{2}}{2} \cdot \dfrac{1}{2} = \dfrac{\sqrt{2}}{4}(\sqrt{3} + 1)$.

Notice that this is the same answer obtained in the last example. $\blacksquare$

If you use the subtraction formulas with $A = \dfrac{\pi}{2}$ and $B = \theta$, you have the following two identities:

$$\boxed{\sin\left(\frac{\pi}{2} - \theta\right) = \cos\theta} \quad \text{and} \quad \boxed{\cos\left(\frac{\pi}{2} - \theta\right) = \sin\theta}.$$

Note that given an angle θ, the complementary angle is $\dfrac{\pi}{2} - \theta$. <u>Co</u>sine is actually the complementary-sine function, since the sine of the angle equals the cosine of the complementary angle, and vice versa. Since $15 = 90 - 75$, $\sin 75°$ therefore equals $\cos 15°$.

<u>Note:</u> You can write the addition and subtraction formulas together as

$$\boxed{\sin(A \pm B) = \sin A \cos B \pm \sin B \cos A}$$

and

$$\boxed{\cos(A \pm B) = \cos A \cos B \mp \sin A \sin B}.$$

Exercises 10.1

1) Write $\cos^7 x$ as $\cos x \cdot (\text{some function of } \sin x)$.

2) Write $\sec^5 x$ as $\sec^2 x \cdot (\text{some function of } \tan x)$.

3) Write $\dfrac{\tan^5 x}{\cos x}$ as $(\sec x \tan x) \cdot (\text{some function of } \sec x)$.

4) Using the sum and difference formulas for sine and cosine, derive the following results

$$\tan(A + B) = \frac{\tan A + \tan B}{1 - \tan A \tan B}, \quad \text{and} \quad \tan(A - B) = \frac{\tan A - \tan B}{1 + \tan A \tan B}.$$

5) Using the results in exercise 4, write $\tan 2A$ in terms of $\tan A$.

6) Using the sum and difference formulas for sine and cosine, derive the following results:

$$\cot(A + B) = \frac{\cot A \cot B - 1}{\cot B + \cot A}, \quad \text{and} \quad \cot(A - B) = \frac{\cot A \cot B + 1}{\cot B - \cot A}.$$

7) Using the results in exercise 6, write $\cot 2A$ in terms of $\cot A$.

8) Show that $\cot\left(\dfrac{\pi}{2} - \theta\right) = \tan \theta$.

9) Show that $\csc\left(\dfrac{\pi}{2} - \theta\right) = \sec \theta$.

Exponential Functions

A.1   Introduction

An exponential function has the form $f(x) = a^x$, where $a > 0$. The number a is called the <u>base</u>. Consider $a = 2$. It is clear what $f(x) = 2^x$ means for some values of x. For example

$$f(0) = 2^0 = 1, \qquad f(1) = 2^1 = 2,$$

$$f(2) = 2^2 = 4, \qquad f(3) = 2^3 = 8,$$

$$f(-1) = 2^{-1} = \frac{1}{2}, \qquad f(-2) = 2^{-2} = \frac{1}{4},$$

$$f(\tfrac{1}{2}) = 2^{\frac{1}{2}} = \sqrt{2} \cong 1.414,$$

and $\qquad f(3.2) = 2^{3.2} = 2^3 \cdot 2^{\frac{1}{5}} = 8\sqrt[5]{2}.$

This last one could be tough to calculate, but at least you know <u>what it means</u>. Use the above values to plot these points for the graph of $y(x) = 2^x$.

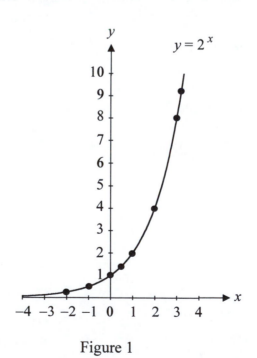

Figure 1

We know what all rational exponents of 2 mean: $2^{\frac{m}{n}} = (\sqrt[n]{2})^m$, if $\frac{m}{n}$ is in lowest terms. What happens at all the remaining (irrational) points? It can be shown (but not here) that there is exactly <u>one</u>

smooth curve, always increasing, that can be drawn through all these points. That is the graph of 2^x, for x real. Notice that its domain is $(-\infty,\infty)$ and the range is $(0,\infty)$.

Graphs of a^x, for $a > 1$

Using these methods, plot the family of graphs a^x. Figure 2 shows several exponential functions for the case when $a > 1$.

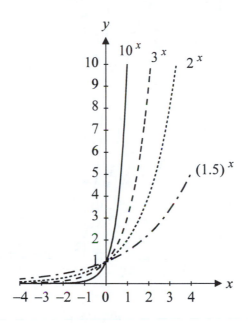

Figure 2

Note: a) As x gets large, each of these functions increases without bound (goes to ∞), but 10^x does it much faster than 3^x, etc.

b) As x goes to $-\infty$, these functions go to 0. But again, 10^x does it much faster than 3^x, etc.

c) All exponential functions pass through the point (0,1).

Graphs of a^x, for $0 < a < 1$

Figure 3 shows several exponential functions for the second case, where $0 < a < 1$, which can be obtained by plotting a few points. (A few thousand that is, if you're the computer plotting this graph.)

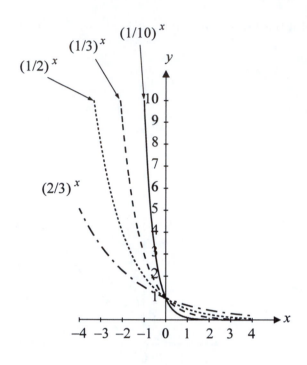

Figure 3

By comparing Figures 2 and 3, it certainly <u>looks</u> like $\left(\dfrac{1}{10}\right)^x$ is the mirror image of 10^x, and $\left(\dfrac{1}{2}\right)^x$ is the mirror image of 2^x, etc. That this is true can be seen by a little computation. Let $f(x) = 2^x$, and let $g(x) = \left(\dfrac{1}{2}\right)^x$. [Remember: $g(x)$ is the mirror image about the y-axis of $f(x)$ iff $g(x) = f(-x)$.] Well,

$g(x) = \left(\dfrac{1}{2}\right)^x = \dfrac{1^x}{2^x} = \dfrac{1}{2^x} = 2^{-x} = f(-x)$. This point is made graphically in Figure 4.

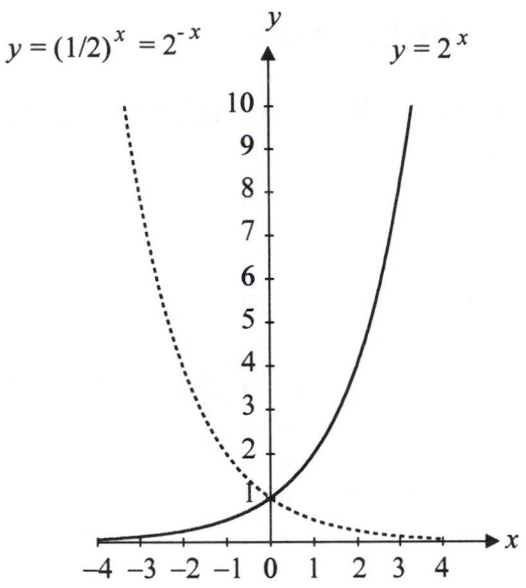

$$y = (1/2)^x = 2^{-x} \qquad\qquad y = 2^x$$

Figure 4

By the way, notice that a^x is defined <u>for all x</u> only if $a > 0$. If $a < 0$, you can no longer have a^x for all x. For example if $a = -1$, and $x = \frac{1}{2}$, we have $(-1)^{\frac{1}{2}} = \sqrt{-1}$, which is not a real number. So the family of functions a^x is defined for $a > 0$ and any real number x.

Exercises A.1

1) Determine the behavior of the following exponential functions as $x \to \pm\infty$, then sketch the graph of the function, labeling at least three points.

   a) $f(x) = \left(\dfrac{2}{3}\right)^x$

   b) $f(x) = \left(\dfrac{3}{2}\right)^x$

   c) $f(x) = 1.1^x$

   d) $f(x) = .32^x$

2) Graph the function $f(x) = 2^x$, then using that result and the methods learned in Chapter 3, graph the following functions:

   a) $f(x) = 2^{x-1}$ b) $f(x) = 2^{x+3}$ c) $g(x) = -2^{x+2}$ d) $f(x) = -2^{x-1}$

3) Graph the function $f(x) = 3^x$, then using that result, graph the following functions:

   a) $f(x) = 3^{x+1}$ b) $f(x) = -\frac{1}{2} \cdot 3^{x+1}$.

A.2 The Function e^x – also called THE Exponential

Of all the exponential functions a^x, the one that has a 45° tangent at $x = 0$ is especially important. This exponential is depicted in Figure 5.

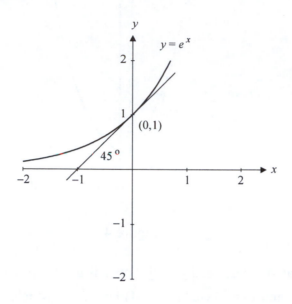

Figure 5

You can tell from the graphs of 2^x and 3^x that this special function lies between them. The particular value of a that gives this exponential is called "e." It can be calculated that $e \cong 2.718$. (In fact, $e \cong 2.718281828459045....$) The exact reason why e^x is so important becomes clear in calculus, when you see that of all a^x, e^x has the simplest derivative.

Since any function a^x can be written as e^{kx}, for some k, that is the form ordinarily used in mathematics, science, and engineering. (See Appendix C for details.)

Exercises A.2

1) Sketch: a) $y = e^{x-1}$ b) $y = e^{-x}$ c) $y = -\frac{1}{2}e^x$

d) $f(x) = e^x - 1$ e) $y = -\frac{1}{2}e^{-x}$

2) Sketch e^{-x^2}. (Hint: Plot and think!)

Appendix B

Inverse Functions

B.1 The Idea of Inverses

When one function undoes the action of another, it is said to be the <u>inverse of the other</u>. For example, look at $f(x) = x^3$ and $g(x) = \sqrt[3]{x}$. If you take any number, x, cube it, and then take the cube root of the result, you're back to x. (Try this for $x = 2$ and 3.) In symbols

$$\sqrt[3]{x^3} = x, \text{ or } g(f(x)) = x.$$

Similarly, you can show $\qquad \left(\sqrt[3]{x}\right)^3 = x, \text{ or } f(g(x)) = x.$

A second example is the doubling function, $f(x) = 2x$, and the halving function, $g(x) = \dfrac{x}{2}$; they're inverse to each other. The function $f(x) = \dfrac{1}{x}$ is its own inverse!

Definition: We say $f(x)$ and $g(x)$ are <u>inverse to each other</u> if $f(g(x)) = x$ and $g(f(x)) = x$, and if domain of f = range of g, and domain of g = range of f.

<u>Note:</u> The inverse of a function f is denoted f^{-1}.

Note that the domains and ranges are important to the discussion of inverse functions. Two function expressions can be inverses over one interval but not inverses over another interval. (See exercise 8 in section B.3.)

Inverses, and finding them, are a big deal in mathematics. Here's just a little example. Suppose you wish to solve the equation $f(x) = 0$. If you could find f^{-1}, you could apply it to both sides to get $f^{-1}(f(x)) = f^{-1}(0)$, and so $x = f^{-1}(0)$. PRESTO! You've solved the equation.

The question is: Okay, now that you've got a function f, how do you find f^{-1}? Specifically: if you've got the graph of f, how do you find the graph of f^{-1}? Or, if you've got an expression for f, how do you get the expression for f^{-1}? Read on!

B.2 Finding the Inverse of f Given by a Graph

First of all, notice that the graph of f is the set of points of the form $(x, f(x))$. (Do you agree?) For every x in the domain of f, f takes x to $f(x)$. Use the following symbols:

$$x \xmapsto{\;\;f\;\;} f(x)$$

Notice that f^{-1} takes $f(x)$ to x. So

$$f(x) \xmapsto{\;\;f^{-1}\;\;} x$$

Hence the graph of f^{-1} is exactly the set of points $(f(x), x)$. (Notice, for example, that the point $(2,8)$ is on the graph of x^3, and $(8,2)$ is on the graph of $\sqrt[3]{x}$.) The result is that the graph of f^{-1} is just the graph of f with the <u>order of the coordinates reversed.</u> How do you do that, you ask? The picture below tells all!

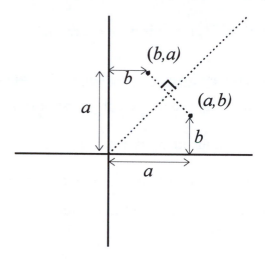

Figure 1

Using geometry, you can prove that (a,b) and (b,a) are reflections of each other about the line $y = x$, i.e., the line through the origin at an angle of $45°$. So to go from the graph of f to the graph of f^{-1}, you can simply reflect the entire graph of f about the line $y = x$.

If $f(x) = 2x$ and $g(x) = \dfrac{x}{2}$, the picture looks like

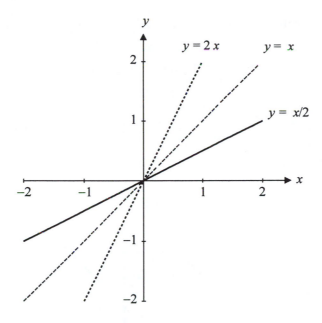

Figure 2

Notice that each graph is obtained from the other by reflecting across the line $y = x$.

Consider the functions $f(x) = x^3$ and $g(x) = \sqrt[3]{x}$. They are also reflections of each other about the line $y = x$.

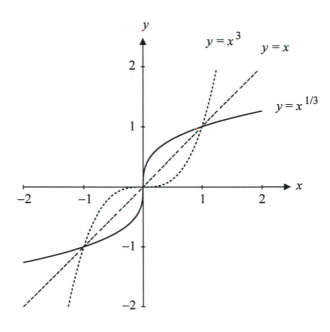

Figure 3

Example 1:   Show that functions $f(x) = x+2$ and $g(x) = x-2$ are inverses of one another on the interval $(-\infty, \infty)$, and then graph the functions.

Solution: Since $f(g(x)) = f(x-2) = (x-2)+2 = x$ and
$g(f(x)) = g(x+2) = (x+2)-2 = x$ for any value of x, these two functions are inverses on the entire interval $(-\infty, \infty)$. The graphs are shown below.

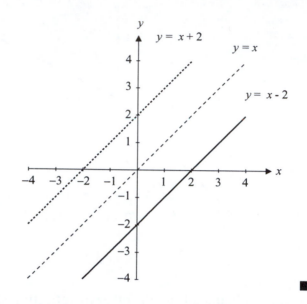

The function $f(x) = \dfrac{1}{x}$ is its own inverse since it reflects back onto itself, as shown in Figure 4.

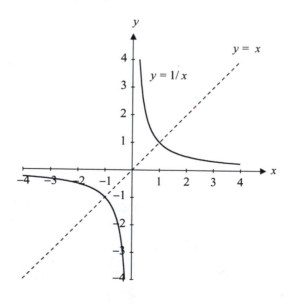

Figure 4

Notice that some functions don't have inverse functions. Here's an example.

Example 2:   Show graphically that the function $f(x) = x^2$ does not have an inverse function on the interval $(-\infty, \infty)$.

Solution: A picture's worth a thousand words.

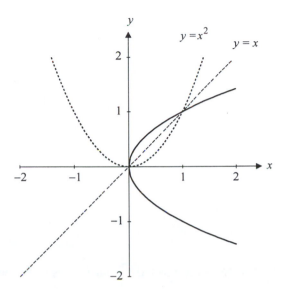

You can flip the graph of $f(x) = x^2$ but the resulting graph (given by the solid curve) is NOT a function. Think about it: how would you evaluate it at $x = 1$, for example? There are two candidates, 1 and -1. A function can have at most one value for each x. In other words, each vertical line can cross the graph of a function at most once. That's called the "<u>vertical line test for being a function</u>." We conclude that $f(x) = x^2$ on $(-\infty, \infty)$ does not have an inverse function. ■

We got into this mess because our original function $f(x) = x^2$ had points where two different x values had the same $f(x)$ value. That is, it was crossed by horizontal lines more than once each. It flunked the "<u>horizontal line test for a function to have an inverse</u>." Any function whose graph is crossed by a given horizontal line at most once is called <u>invertible</u> – that is, f^{-1} exists. Another term for invertible is <u>one-to-one</u>, written 1-1. (Why is this a good name?) So the problem was that $f(x) = x^2$ is not 1-1. But suppose you really wanted an inverse for $f(x) = x^2$. You could (a) get depressed or (b) settle for a piece of the whole thing, as in the next example.

Example 3:   Consider the function $f(x) = x^2$, for $x \geq 0$. (Notice the restricted domain.) Graph its inverse.

Solution: No problem!

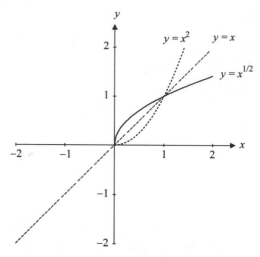

All the tests are passed. That inverse, by the way, is $g(x) = x^{\frac{1}{2}} = \sqrt{x}$. ∎

Example 4:   Consider $f(x) = \sin x$, for $-\dfrac{\pi}{2} \leq x \leq \dfrac{\pi}{2}$. Graph its inverse.

Solution: By flipping this restricted sine function about the line $y = x$, you'll get its inverse, which is called inverse sine and also arcsine. It is denoted $\arcsin x$ or $\sin^{-1} x$.

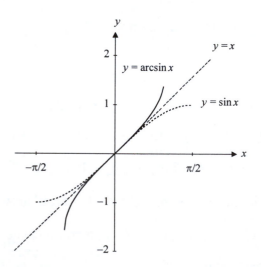

You'll see more of this . . . .   ∎

Exercises B.2 Graph the following, and their inverses, if they exist. If they do not, explain why.

1) $f(x) = 2x - 1$

2) $f(x) = x^5$

3) $g(x) = -x$

4) $k(y) = \cos y$

5) $h(x) = 2^x$

6) $L(t) = 1^t$

7) $f(x) = \left(\dfrac{1}{2}\right)^x$

8) $f(x) = x^3 - 1$

9)   Is the function $f(x) = \cos x$ invertible for $0 \le x \le \pi$? How about for the interval
$-\dfrac{\pi}{2} \le x \le \dfrac{\pi}{2}$? (Sketch!)

B.3 Finding the Inverse of *f* Given by an Expression

Consider the equation $y = f(x)$. **If f^{-1}** exists, you can apply it to both sides and get $f^{-1}(y) = x$. You thus have f^{-1} expressed in the variable y. Merely replace it with x and you're done. Consider the following examples.

Example 1:   Find the inverse of the function $f(x) = \dfrac{2}{x+3}$.

Solution: Write $y = \dfrac{2}{x+3}$ and solve this equation for x to obtain $f^{-1}(y)$. Multiplying by $x + 3$ gives

$$x\,y + 3\,y = 2,$$ which in turn means

$$x\,y = 2 - 3y,$$ and hence

$$x = \frac{2 - 3y}{y} = f^{-1}(y).$$ Using x instead of y gives

$$f^{-1}(x) = \frac{2 - 3x}{x}. \blacksquare$$

You can check this example by determining whether $f\!\left(f^{-1}(x)\right) = x$ and $f^{-1}\!\left(f(x)\right) = x$. In this case

$$f^{-1}\!\left(f(x)\right) = f^{-1}\!\left(\frac{2}{x+3}\right)$$

$$= \frac{2 - 3\left(\dfrac{2}{x+3}\right)}{\dfrac{2}{x+3}}$$

$$= \frac{\dfrac{2x + 6 - 6}{x+3}}{\dfrac{2}{x+3}} = \frac{\dfrac{2x}{x+3}}{\dfrac{2}{x+3}} = \frac{2x}{2} = x.$$

Also

$$f\!\left(f^{-1}(x)\right) = f\!\left(\frac{2 - 3x}{x}\right)$$

$$= \frac{2}{\dfrac{2-3x}{x} + 3}$$

$$= \frac{2}{\dfrac{2-3x+3x}{x}} = \frac{2}{\dfrac{2}{x}} = x.$$

Check! We're done.

Example 2:   Find the inverse of $f(x) = \sqrt[3]{2x+1}$.

Solution: Write $y = \sqrt[3]{2x+1}$, and solve for x.

$$y^3 = 2x + 1$$

$$2x = y^3 - 1$$

$$x = \frac{y^3 - 1}{2} = f^{-1}(y),$$

$$\therefore f^{-1}(x) = \frac{x^3 - 1}{2}. \quad \blacksquare$$

Question: It all looks so easy. Can anything go wrong? And what if a given function doesn't <u>have</u> an inverse? How will that show up?

Answer: Yes, things can deteriorate. For example, what if you can't solve for x? If that happens, then you're stuck. Moreover, if f^{-1} doesn't exist, it will show up by the fact that the equation <u>cannot</u> be solved for x. Not even by Gauss, with help from Einstein. See the next example.

Example 3:   Find the inverse of $f(x) = x^4 - 3$.

Solution: Write $f(x) = x^4 - 3$, and solve for x.

$$x^4 = y + 3$$

$$x = \pm \sqrt[4]{y+3}$$

Aha, you see, that's not a function. You have not solved for x as a function of y. Of course, you knew that $f(x)$ does not have an inverse, because it flunks the horizontal line test. ∎

Exercises B.3   Find the expressions of the inverses if they exist. If they do not, explain why.

1) $f(x) = 2x - 3$ 2) $k(x) = \dfrac{x}{x+1}$ 3) $g(x) = \sqrt[3]{5x + 1}$

4) $s(t) = \sqrt{t + 2}$ 5) $f(x) = \dfrac{2}{x}$ 6) $f(w) = \dfrac{w^2}{w^2 + 1}$

7)   The function $f(x) = (x-1)^4$ does not have an inverse on the interval $(-\infty, \infty)$. Show this. Then show that, if you restrict the domain to $[1, \infty)$, this restricted function has an inverse. Graph both functions.

8)   a) Let $f(x) = x^2 + 3$ on $[0,1]$. Find its domain and range and then sketch it and its inverse. Find an expression for the inverse $f^{-1}(x)$.

b) Let $g(x) = x^2 + 3$ on $[-1,0]$. Do the same as in part (a).

c) The function expressions of f and g are the same. Are their inverses the same?

Appendix C

Logarithmic Functions

C.1 Definition of Logarithms

There's one more function you'll meet in science, engineering, and economics, and sometimes in the newspaper. In symbols, it looks like this:

$$\log_a x$$

It is read as "log, to the base a, of x." (Do NOT read this as "the log of a-to-the-x"!) There are several ways of defining it. Here's one.

Definition: Let $a > 0$, $a \neq 1$. Then $\log_a x$ is <u>the number to which you raise a to get x.</u>

Example 1: Demonstrate that $\log_2 8 = 3$.

Solution: Here the base is 2 and $x = 8$. To what number do you have to raise 2 in order to get 8? Answer: 3, so $\log_2 8 = 3$. ■

Example 2: Show that $\log_{10} 1{,}000{,}000 = 6$.

Solution: Here the base is 10, and $x = 1{,}000{,}000$. What number do you have to raise 10 to, in order to get 1,000,000 (6 zeros)? Answer: 6, so $\log_{10} 1{,}000{,}000 = 6$. ■

Example 3: $\log_{10} .01 = ?$

Solution: Here the base is 10, and $x = .01$. Write .01 as a power of 10. Here you go:

$$.01 = \frac{1}{100} = \frac{1}{10^2} = 10^{-2}$$

What number do you have to raise 10 to in order to get 10^{-2}? Answer: −2, of course, so $\log_{10} .01 = -2$. ■

Example 4: $\log_2 32 = ?$

Solution: Write 32 as a power of 2:

$$32 = 2 \cdot 2 \cdot 2 \cdot 2 \cdot 2 = 2^5,$$

so, $\log_2 32 = 5.$ ∎

Example 5: $\log_3 81 = ?$

Solution: Write 81 as a power of 3:

$$81 = 3 \cdot 3 \cdot 3 \cdot 3 = 3^4,$$

so, $\log_3 81 = 4.$ ∎

Example 6: $\log_7 7^{15} = ?$

Solution: 7^{15} is already written as a power of 7. What a silly question! $\log_7 7^{15} = 15.$ ∎

Example 7: $\log_a a^3 = ?$

Solution: Another silly question! If you write a^3 as a power of a, obviously the exponent must be 3, so $\log_a a^3 = 3.$ ∎

Remark: Instead of writing $\log_{10} x$, we often don't bother writing the 10, and write $\log x$. This is called the common logarithm.

Exercises C.1

Evaluate the following:

1) $\log_9 81$ 2) $\log_2 \dfrac{1}{2}$ 3) $\log \dfrac{1}{1000}$ 4) $\log_{\frac{1}{2}} 8$

5) $\log_3 \sqrt{3}$ 6) $\log_4 2\sqrt{2}$ 7) $\log_b b^{13}$ 8) $\log_a a^x$

C.2 Logs as Inverses of Exponential Functions; Graphs and Equations

You will see that there is a fundamental relationship between logs and exponentials: they are inverse to each other. Recall from Appendix B, that f and g are called inverse to each other if all the following are true:

a) $f(g(x)) = x$

b) $g(f(x)) = x$

c) domain of f = range of g

d) domain of g = range of f

Theorem: Let $a > 0$, $a \neq 1$. Then $\log_a x$ and a^x are inverse to each other.

Remark: This theorem has been proved rigorously using the continuity properties from calculus, but that cannot be done here. However, if you let $f(x) = a^x$ and $g(x) = \log_a x$, you can examine the first two conditions.

a) $f(g(x)) = a^{g(x)} = a^{\log_a x}$. What does this mean? It is a, raised to the power to which you raise a to get x. So it equals x – i.e., $f(g(x)) = a^{g(x)} = a^{\log_a x} = x$.

b) $g(f(x)) = \log_a f(x) = \log_a a^x = x$. (Just like examples 6 and 7 in the previous section!)

So you haven't completely proved this theorem, but you can see that $\log_a x$ and a^x undo each other.

Knowing that $\log_a x$ and a^x are inverses allows you immediately to graph $\log_a x$. If you wish to graph the function $f(x) = \log_2 x$, you need only graph the function $g(x) = 2^x$, and flip it around the line $y = x$ (see Figure 1).

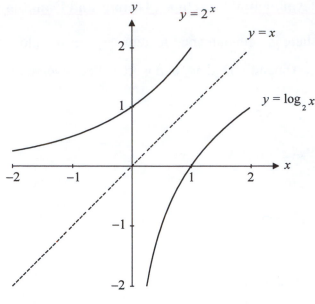

Figure 1

Notice that the domain of $f(x) = \log_2 x$ is the set of all positive numbers, and the range is the set of all numbers. Notice also $\log_2 1 = 0$. Figure 2 shows the common logarithm.

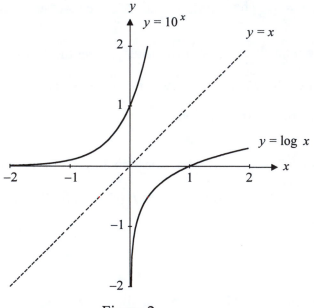

Figure 2

The next example shows the logarithm for base $a = \dfrac{1}{2}$. First graph the function $\left(\dfrac{1}{2}\right)^x$, and then flip it about the line $y = x$ to obtain the graph of $f(x) = \log_{\frac{1}{2}} x$.

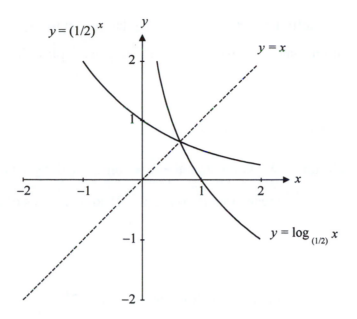

Figure 3

Typically, you'll come across $\log_a x$ only for $a > 1$. Note that $\log_{1/a} x = -\log_a x$. You could get the previous graph by first graphing $\log_2 x$, and then rotating it about the x-axis! As with other functions, once you know what the log graph looks like, you can obtain other graphs by shifting and stretching. Figure 4 shows how the graph of $\log_2 (x+3)$ is obtained from $\log_2 x$ by shifting it to the left 3 units.

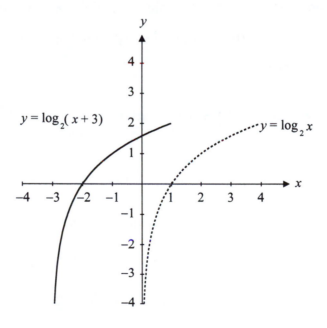

Figure 4

By recalling the definition of logs and their relationship to exponentials, it is possible to solve some new equations containing such animals. The following examples show some of the problems you may encounter.

Example 1:   Solve $\log_2 x = 4$.

  Solution: We know that the function 2^x <u>undoes</u> the action of that log function. This means that if you apply 2^x to both sides of the equation – that is,

$$2^{\log_2 x} = 2^4,$$

the left side becomes x, so $x = 2^4 = 16$.

(The left side is x, because the log was "undone" by the action of 2^x; this is the true meaning of "inverseness," which is how $\log_2 x$ and 2^x are related.) ∎

Example 2:   Solve $\log_{10} x = 3$.

  Solution: As in the last example, to "undo" the action of the log (now with base 10) apply 10^x to both sides. (Recall that applying 10^x to 3 means evaluating 10^x at 3.)

So $x = 10^3 = 1000$. ∎

Example 3:   Solve $\log_{10} (x^2 - 4x + 14) = 1$.

  Solution: Again, apply 10^x to both sides. So,

$$(x^2 - 4x + 14) = 10^1 = 10,$$

or $x^2 - 4x + 4 = 0,$

which factors as $(x - 2)^2 = 0$. So $x = 2$. ∎

Example 4:   Solve $\log_3 (x^2 - 3x - 7) = 1$.

Solution: To "peel off" the log, apply 3^x, and in doing so you get

$$(x^2 - 3x - 7) = 3^1 = 3,$$

or $x^2 - 3x - 10 = 0.$

Factor this last equation to give

$$(x - 5)(x + 2) = 0,$$

which has solutions $x = 5$ and $x = -2$. ∎

Example 5:   Solve $2^{x^2 + 1} = 8$.

Solution: Apply $\log_2 x$ to both sides (i.e., take $\log_2$ of both sides) and get

$$x^2 + 1 = \log_2 8,$$

or $x^2 + 1 = 3.$

This gives

$$x^2 = 2,$$

or $x = \pm \sqrt{2}$. ∎

Example 6:   Solve $100^{\sin x} = 10$, for $x \in \left[0, \dfrac{\pi}{2} \right]$.

Solution: Apply $\log_{100} x$ to both sides.

Therefore $\sin x = \log_{100} 10,$

and since $\log_{100} 10 = \tfrac{1}{2},$

we have $\sin x = \tfrac{1}{2},$

or $x = \dfrac{\pi}{6}$. ∎

Exercises C.2

1) a) Graph $f(x) = 2^{x+1}$.

   b) Show that $g(x) = \log_2 \dfrac{x}{2}$ is the inverse of $f(x)$ and graph it.

2) Using a calculator, graph $y = \log x$ between .5 and 2 by using several x values. Using the same x values, graph $y = \log(10x)$. What do you notice?

3) Solve the following:

   a) $2^{x-3} = 64$ b) $3^{x+1} = 27$ c) $4^{2x-3} = 16$ d) $5^{x+5} = \dfrac{1}{125}$

4) Solve $\log_3 (x+7) = -1$.

5) Solve $\log_{64} x^2 = \dfrac{1}{3}$.

6) Solve $\log_3 (x^2 - 5) = 2$.

7) Solve $\log(\cos x) = 0$ for $x \in [0, 2\pi]$.

C.3 Laws of Logarithms

Your life among the logs is made much simpler when you know certain log laws. They're great when you're solving equations or simplifying expressions.

Log Laws:

1) $\log_a xy = \log_a x + \log_a y$

2) $\log_a \dfrac{x}{y} = \log_a x - \log_a y$

3) $\log_a x^r = r \log_a x$

4) $\log_a 1 = 0$ for all a (but we knew that already!)

5)   For any $a > 0$, with $a \neq 1$, $\log_a x = \dfrac{\log_b x}{\log_b a}$, for any convenient b – for example, 10 or e.

This "change of base" law can be a lifesaver if you can't handle $\log_a x$.

Example 1:   Solve $\log_2 x^2 + \log_2 2x = 4$.

Solution: You can combine the left side using rule 1.

So $\log_2 2x^3 = 4$,

using the first law.

Now you can apply 2^x (before you couldn't):

$$2x^3 = 2^4 = 16$$

$$x^3 = 8$$

$$x = 2 .$$

You should check this solution:

$$\log_2 (2^2) + \log_2 (2 \cdot 2) = \log_2 4 + \log_2 4 = 2 + 2 = 4 . \quad \blacksquare$$

Example 2:   Solve $\log_{10}(x^2 - 3x)^3 = 3$.

Solution: Using the third law, you obtain

$$3\log_{10}(x^2 - 3x) = 3, \text{ or}$$

$\log_{10}(x^2 - 3x) = 1$. Now apply 10^x:

$$x^2 - 3x = 10^1 = 10.$$

Hence $x^2 - 3x - 10 = 0,$

$$(x - 5)(x + 2) = 0,$$

and $x = 5 \text{ or } -2.$ ■

In this example, you calculated two answers, both of which were valid because they satisfied the original equation. Sometimes, however, you may get some answers that are not valid. (They are called extraneous solutions.) You must check your solutions to determine that they are valid. Consider the following.

Example 3:   Solve $\log_2 x + \log_2(x-1) = 1$.

Solution: Using the first law, you obtain

$$\log_2 x + \log_2(x-1) = \log_2\big(x(x-1)\big) \text{ which gives}$$

$\log_2\big(x(x-1)\big) = 1$. Now apply 2^x and get

$$x(x-1) = 2.$$

Hence $x^2 - x - 2 = 0$

and $(x-2)(x+1) = 0$, and

$$x = 2 \text{ or } -1.$$

However, when checking these "solutions," notice that the initial equation is not satisfied for $x = -1$, so $x = 2$ is the only solution.   ■

Example 4:   Evaluate $\log_7 5$.

Solution: You can't evaluate this directly. You could use a calculator, but it doesn't have $\log_7 x$. But

$$\log_7 5 = \frac{\log_{10} 5}{\log_{10} 7},$$

so $$\log_7 5 \cong \frac{.699}{.845} \cong .827. \quad \blacksquare$$

Exercises C.3

1) Solve: $\log_3 x + \log_3 (x-6) = 3$.

2) Solve: $\log y + \log y^2 = -1$.

3) Solve: $\log_6 (2x + 1) - \log_6 (2x - 1) = 1$.

4) Solve: $\log_2 x^2 - \log_2 (3x - 8) = 2$.

5) Find numbers a, x, and y where the value of $\log_a (x + y)$ is not equal to the value of $\log_a x + \log_a y$. (<u>Hint:</u> There are many such numbers.)

6) Approximate $\log_3 4$ using the change of base formula and your calculator.

C.4 The Natural Logarithm

In Appendix A, we introduced the exponential function e^x. Its inverse, $\log_e x$, is called the <u>natural logarithm</u>. For simplicity, $\log_e x$ is denoted by the symbol $\ln x$. To get its graph, we flip e^x about the line $y = x$.

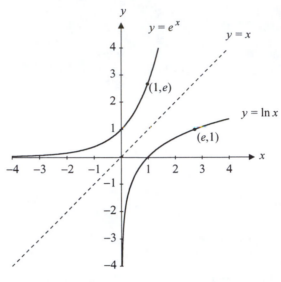

Figure 5

Notice that $\ln 1 = 0$, and $\ln e = 1$. You will also encounter the natural logarithm when you study integration. It helps to know about $\ln x$, and how to graph related functions. Figure 6 shows several horizontal shifts of the natural logarithm.

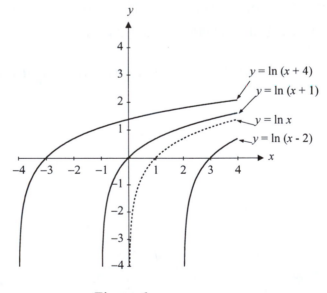

Figure 6

In solving equations, you handle the natural log the same as any other log, remembering that the base is that special number e.

Example 1:   Solve $\ln(x^2 - 1)^3 = 1$.

Solution: Apply e^x to get

$$(x^2 - 1)^3 = e,$$

or $$(x^2 - 1) = e^{\frac{1}{3}},$$

which can be solved to give

$$x = \pm\sqrt{1 + e^{\frac{1}{3}}}.$$

For most purposes you can leave your answer in this form. If its decimal approximation is needed, you can use your calculator to determine one. ∎

Example 2:   Solve $e^{\sin x} = 1$, $x \in [0, 2\pi]$.

Solution: Apply $\ln x$ to get

$$\sin x = \ln 1 = 0,$$

and so $$x = 0, \pi, 2\pi. \quad \blacksquare$$

Remark: As mentioned in Appendix A, any exponential a^x can be written in the form e^{kx} for some constant k. <u>The TRICK:</u> write a as $e^{\ln a}$. So

$$a^x = (e^{\ln a})^x = e^{(\ln a)x} \quad \text{(Recall: } (a^m)^n = a^{mn}.\text{)}$$

That's it! Now consider the following:

Example 3:   Write 2^x in the form e^{kx}.

Solution: $2^x = (e^{\ln 2})^x = e^{(\ln 2)x}$,

and since $\ln 2 \cong .693$,

$$2^x \cong e^{.693x}. \quad \blacksquare$$

Example 4:   Write $(.345)^x$ in the form e^{kx}.

Solution: $(.345)^x = (e^{\ln .345})^x = e^{(\ln .345)x}$,

and since $\ln .345 \cong -1.06$,

$(.345)^x \cong e^{-1.06x}$.   ■

Remarks: a) If you convert the function a^x into the form e^{kx}, then if $a > 1$, as in Example 3, the constant $k > 0$. All functions of the form e^{kx}, for $k > 0$, look similar to the graph shown in Figure 7.

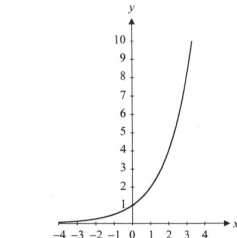

Figure 7

In this case, the function is said to have <u>exponential growth</u>. An example from biology would be the growth of a colony of bacteria under ideal conditions.

b) On the other hand, if $0 < a < 1$, as in Example 4, then the constant $k < 0$. All functions of the form e^{kx}, for $k < 0$, look similar to the graph shown in Figure 8.

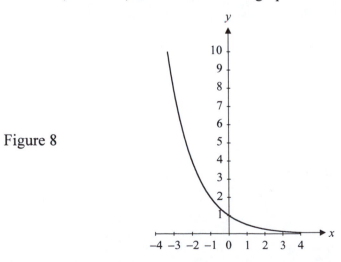

Figure 8

In this case, the function is said to have <u>exponential decay</u>. In physics, the decay of radioactive substances is represented by such functions.

Exercises C.4

1)   Graph e^{x-1}.

2)   Graph $y = \ln(x-1)$.

3)   Graph $y = 2 + \ln(x+1)$.

4)   Solve $\ln t + \ln t^2 = 6$.

5)   Solve $e^{x^2 + 2x - 3} = 1$.

6)   Solve $e^{\ln(w^2+1)} = 5$. ($\leftarrow$ Gift!)

7)   Write 10^x and $(0.5)^x$ in the form e^{kx}. (Use a calculator.)

Appendix D

The Binomial Theorem

A binomial is the sum of two terms and can therefore be represented as $a + b$. The binomial theorem states what happens when powers of $a + b$ are multiplied out. Let's expand $(a + b)^n$ for $n = 1, 2, 3, 4$.

$$(a + b)^1 = a + b$$

$$(a + b)^2 = a^2 + 2ab + b^2$$

$$(a + b)^3 = a^3 + 3a^2b + 3ab^2 + b^3$$

$$(a + b)^4 = a^4 + 4a^3b + 6a^2b^2 + 4ab^3 + b^4$$

We see some obvious patterns. The powers of a decrease as you go from left to right, starting with a^n and ending with a^0 (which equals 1, so it doesn't appear). The powers of b go in reverse, from $b^0 = 1$ to b^n. But what about the coefficients? Let's write them in the form known as Pascal's triangle:

$$
\begin{array}{ccccccc}
 & & & 1 & 1 & & \\
 & & 1 & 2 & 1 & & \\
 & 1 & 3 & 3 & 1 & & \\
1 & 4 & 6 & 4 & 1 & & \\
\end{array}
$$

Each line begins with 1, followed by the line number, n. Also we see that each entry of a particular line is formed by adding the two entries diagonally above it. For example, in the fourth line, 4 is the sum of $1 + 3$, 6 is the sum of $3 + 3$, etc. To get the next line, notice that $1 + 4 = 5$, $4 + 6 = 10$, etc. – which means that the next line is 1 5 10 10 5 1.

Binomial Theorem: Let n be a positive whole number, and let a and b be any numbers. Then

$$(a + b)^n = ()a^n + ()a^{n-1}b + ()a^{n-2}b^2 + \cdots + ()a^2b^{n-2} + ()ab^{n-1} + ()b^n,$$

where the coefficients in the empty brackets are given by the entries of the nth line of Pascal's triangle.

Example 1:   Expand $(a + b)^6$.

Solution: Check Pascal's triangle. The fifth and sixth lines are

$$1 \quad 5 \quad 10 \quad 10 \quad 5 \quad 1$$

$$1 \quad 6 \quad 15 \quad 20 \quad 15 \quad 6 \quad 1$$

Hence $(a + b)^6$ is equal to

$$a^6 + 6a^5 b + 15a^4 b^2 + 20a^3 b^3 + 15a^2 b^4 + 6ab^5 + b^6. \quad \blacksquare$$

Example 2:   Expand $(x + 2)^5$.

Solution: Check Pascal's triangle. The fifth line is

$$1 \quad 5 \quad 10 \quad 10 \quad 5 \quad 1.$$

Hence $(x + 2)^5$ is equal to

$$(1) x^5 + (5) x^4 2 + (10) x^3 2^2 + (10) x^2 2^3 + (5) x 2^4 + (1) 2^5,$$

which can be simplified to give

$$(x + 2)^5 = x^5 + 10x^4 + 40x^3 + 80x^2 + 80x + 32. \quad \blacksquare$$

Example 3:   Expand $(a - 3)^4$.

Solution: $(a - 3)^4 = (a + (-3))^4$ (Tricky!)

$$= a^4 + (4) a^3 (-3) + (6) a^2 (-3)^2 + (4) a(-3)^3 + (1)(-3)^4$$

$$= a^4 - 12a^3 + 54a^2 - 108a + 81. \blacksquare$$

Remarks: a) This method is clearly simpler than multiplying $a - 3$ by itself repeatedly, but it can become cumbersome if the number n is too big.

b) The coefficients, which appear in Pascal's triangle, can also be expressed by a formula. This can be found in calculus books in the section on binomial series.

Example 4:   Simplify $\dfrac{(x+h)^4 - x^4}{h}$.

Solution: Check Pascal's triangle. The fourth line is

$$1 \quad 4 \quad 6 \quad 4 \quad 1.$$

Hence

$$(x+h)^4 = (1)x^4 + (4)x^3 h + (6)x^2 h^2 + (4)x h^3 + (1)h^4$$

$$= x^4 + 4x^3 h + 6x^2 h^2 + 4x h^3 + h^4,$$

and $\dfrac{(x+h)^4 - x^4}{h} = \dfrac{4x^3 h + 6x^2 h^2 + 4x h^3 + h^4}{h}$.

$$= 4x^3 + 6x^2 h + 4x h^2 + h^3. \quad \blacksquare$$

Remark: If we expand $(x+h)^n$, we get $x^n + nx^{n-1}h + (\quad)h^2 + (\quad)h^3 + \cdots + h^n$,   where the open brackets contain polynomials in x only, without any factors of h. This fact will be crucial when finding the derivative of x^n.

Exercises D.1

1) Expand $(x+1)^6$.

2) Expand $(y-1)^6$.

3) Expand $(2z+3)^4$.

4) Expand $(x+\Delta x)^5$.

5) Using the result in exercise 4, simplify $\dfrac{(x+\Delta x)^5 - x^5}{\Delta x}$.

Answers to Exercises
Chapter 1

1.1 page 3

1) -19 2) 46 3) $4xy - x + 2y$

4) 0 5) $3xy - 6x - xy^2 + 2y$ 6) $2xz - 3yz$

7) $x^2 y + xy^3 - 4x^2 y^2 + 2y^3$

8) $x^3 y^2 - xy^4 - 8x^2 y^2 + 6x^2 y^3 + 2xy^2$

9) a) 27 in 8 operations   b) 27 in 11 operations, so method 1 is "cheaper."

1.2 page 7

1) $\dfrac{1}{4}$ 2) $\dfrac{\pi^2}{2}$ 3) $\dfrac{3}{5}$ 4) $\dfrac{1}{6}$ 5) $\dfrac{-28}{51}$ 6) $\dfrac{9}{14}$

7) $\dfrac{21y + 14}{3y}$ 8) $\dfrac{x - xy + 2 - 2y}{(1 + y)x}$ 9) $\dfrac{y}{y - 2}$ 10) $\dfrac{w}{xy}$

11) $\dfrac{(x + y)^2}{x}$ 12) $\dfrac{1}{y(x - y)}$

1.3 page 10

1) $\dfrac{7}{12}$ 2) $\dfrac{11}{8}$ 3) $\dfrac{7}{30}$ 4) $\dfrac{5}{16}$ 5) $\dfrac{37}{30}$

6) $\dfrac{-16}{9}$ 7) $\dfrac{22}{45}$ 8) $\dfrac{-25}{66}$ 9) $\dfrac{29}{42}$ 10) $\dfrac{x+y}{xy}$

11) $\dfrac{x-y}{xy}$ 12) $\dfrac{4yz - 2xz + xy}{xyz}$ 13) $\dfrac{yz - z(x + 1) + y(x - 2)}{xyz}$

14) $\dfrac{x(z - xy)}{y(x - z)}$ 15) $\dfrac{w - st}{s - 2tw}$ 16) $\dfrac{y^2 - x^2}{xy}$

17) $\dfrac{4y^3 z^2 - 2xz^2 + xy}{x^2 y^2 z}$ 18) $\dfrac{xy^2 z - x^3 yz + 2x^3 + 2xy^2 - 2x^2 y - 2y^3}{2x^3 yz + 2xy^3 z}$

1.4 page 13

1) $\dfrac{225}{16}$ 2) $\dfrac{9}{2}$ 3) $\dfrac{60}{19}$ 4) $\dfrac{4}{81}$ 5) $32\frac{1}{2}$

6) yz^8 7) $\dfrac{1}{4}$ 8) x^{68} 9) x^{-68}

10) Try $x = y = 1$. L.H.S. $= 2^{30}$, while R.H.S. $= 2$.   11) x 12) $y^{-12} = \dfrac{1}{y^{12}}$

13) $\dfrac{x^2}{y}$ 14) $\dfrac{x^6 z - x^2}{y^5 z^5}$

1.5 page 15

1) 12 2) −4 3) $\dfrac{1}{3}$ 4) −2 5) $\dfrac{2}{7}$

6) $\dfrac{2}{3}$ 7) 32 8) −32 9) 4 10) −4

11) $\dfrac{27}{64}$ 12) 1000 13) .0016 14) $2^{34/15}$ 15) 256

16) $3^{1/2}$ 17) $2^{-13/14}$ 18) $3^6 = 729$ 19) $2^{-1/15}$ 20) $3 + 2 = 5$

1.6 page 17

1) 50 2) .2343 3) 1304.4 4) a) \$12.75 b) \$1124.25
5) \$34.00 6) \$191.20 7) \$240.00 8) \$52,500.00

1.7 page 19

1) a) 3.83×10^5 b) -7.24×10^{-4} c) 3.00 d) 2.00×10^2
2) a) 9.48×10^7 b) -3.09×10^5 c) 3.50×10^{-100} d) 7.66×10^{-2} e) -1.68×10^{-8}
   f) 5.57×10^{-2}

Chapter 2

2.1 page 24

1) a) $f(x) = (x-3)^2 + 6$ b) $h(y) = \left(y + \dfrac{5}{2}\right)^2 - \dfrac{25}{4}$ c) $g(s) = (s+1)^2 - 9$

d) $k(x) = 2\left(x - \dfrac{1}{2}\right)^2 + \dfrac{9}{2}$ e) $f(x) = 3\left(x - \dfrac{7}{6}\right)^2 - \dfrac{37}{12}$ f) $w(x) = \pi\left(x + \dfrac{1}{\pi}\right)^2 - \dfrac{1}{\pi}$

2) a) $\left(x - \dfrac{3}{2}\right)^2 - \dfrac{77}{4} = 0$ b) $-3(x+1)^2 + 18 = 0$ alternatively, $(x+1)^2 - 6 = 0$

3) a) $\left(x + \dfrac{3}{2}\right)^2 + 2(y-2)^2 = \dfrac{41}{4}$ b) $3(x+1)^2 - 2(y+2)^2 = -16$

c) $-(x-2)^2 + (y-8)^2 = 100$ d) $9(x-2)^2 + 4(y+1)^2 = 40$

e) $(x-3)^2 + (y+5)^2 = 0$

Chapter 3

3.1 page 27

1) a) 12   b) 33   c) $(x+h)^3 + 2(x+h)$ d) $8x^3 + 4x$

   e) $-x^3 - 2x$ f) $(2 + \Delta x)^3 + 4 + 2\Delta x$

2) $|0| = 0$, $|\pm 1| = 1$, $|\pm 2| = 2$, etc. so

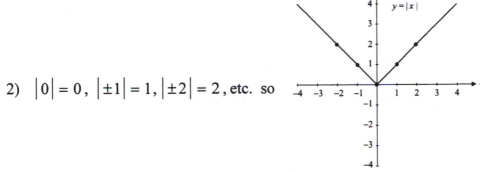

3) a) 1   b) 1   c) 0   d) 1   e) -1

4) a) -1 b) $\pi^4 + 1$ c) $(x+h)^4 - \cos(x+h)$ d) $\dfrac{\pi^4}{16}$ e) $x^4 - \cos x$

5) If the graph crosses or touches the x-axis, then the function is zero there. Hence the solution of the problem is all integer multiples of π, i.e., 0, $\pm \pi$, $\pm 2\pi$, $\pm 3\pi$ ....

6) A function is positive wherever its graph is above the x-axis, while the function is negative wherever its graph is below the x-axis. The solution of this problem amounts to finding points where the graph either touches the x-axis or is above the x-axis. By examining the graph, we obtain $\left[0, \dfrac{\pi}{2}\right] \cup \left[\dfrac{3\pi}{2}, 2\pi\right]$ as the set where the equation is satisfied.

7) All x such that $-2 < x < 2$.

8) All x that are less than -3 or greater than 3, i.e., the interval $(-\infty, -3) \cup (3, \infty)$.

3.2 page 29

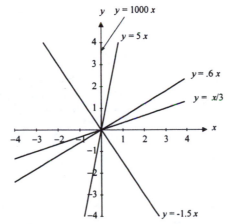

Believe it or not the graph for $y = 1000x$ is really plotted on this graph, but because of the scale you can't really see it. If we scale this as in the next figure you see this but lose the others.

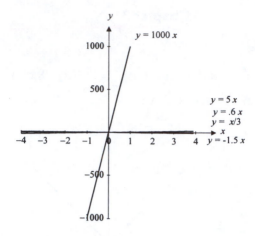

3.3 page 34

1) Our four points are given in the table and plotted on the graph. By picking more and more points, we can "fill in" the picture, getting the solid line in the figure below.

x	$\sqrt{x}$
0	0
1	1
4	2
9	3

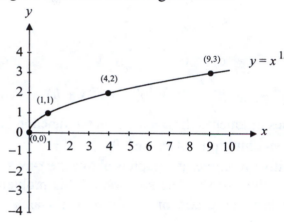

Exercises 2 through 6 and exercise 8 are all shown in the section body.

7) First we notice that the function $x^{2/3}$ is defined for both positive and negative values of x. Some values are given in the following table, and the graph is shown in the following figure.

x	$x^{2/3}$
0	0
1	1
−1	1
8	4

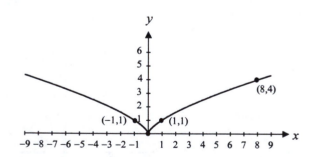

3.4 page 36

1)

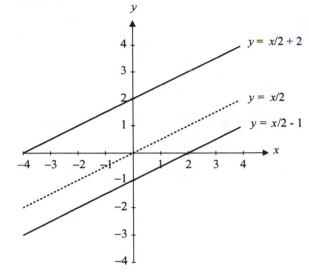

2)

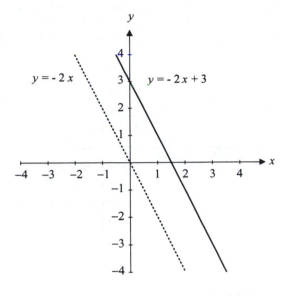

3a),b)

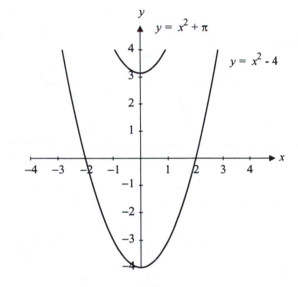

3c)

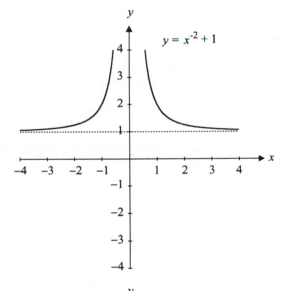

4)

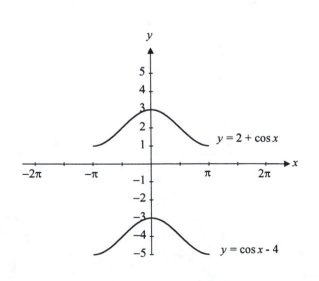

5)

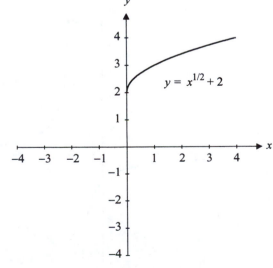

162

6)

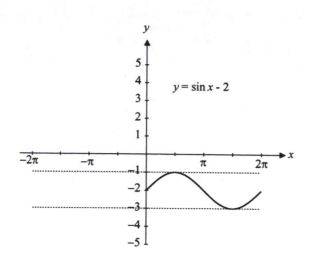

$y = \sin x - 2$

7)

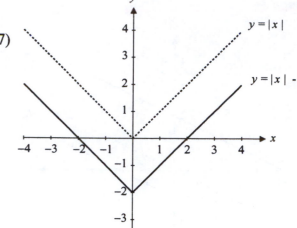

$y = |x|$

$y = |x| - 2$

3.5 page 39

1)

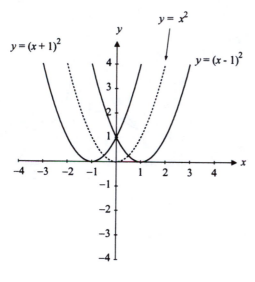

$y = x^2$
$y = (x+1)^2$
$y = (x-1)^2$

2a),b)

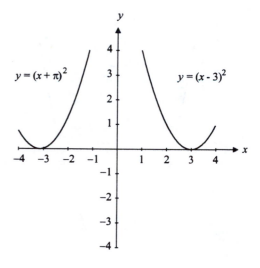

$y = (x+\pi)^2$
$y = (x-3)^2$

2c)

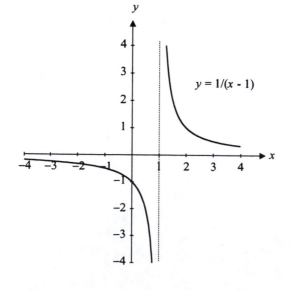

$y = 1/(x-1)$

3a)

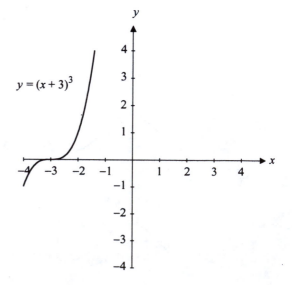

$y = (x+3)^3$

3b)

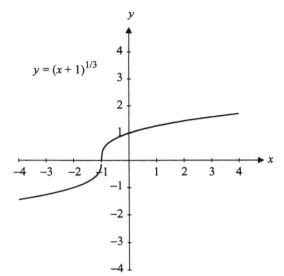

$y = (x+1)^{1/3}$

3c)

$y = -1/(x-a)^4$

where $a > 0$

4)

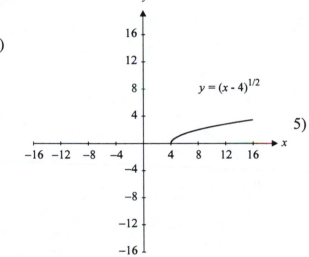

$y = (x-4)^{1/2}$

5)

$y = \cos(x - \pi)$

3.6 page 42

1)

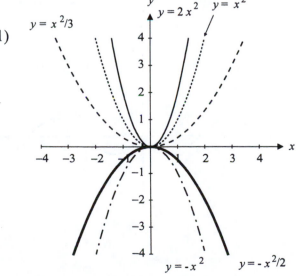

$y = x^2/3$

$y = 2x^2$

$y = x^2$

$y = -x^2$

$y = -x^2/2$

2)

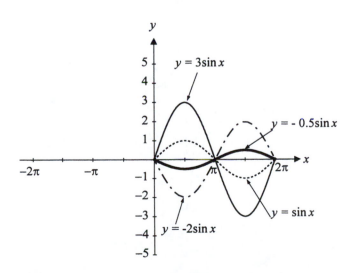

$y = 3\sin x$

$y = -0.5\sin x$

$y = \sin x$

$y = -2\sin x$

3)

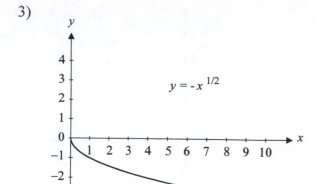

$y = -x^{1/2}$

3.7 page 45

1)

2)

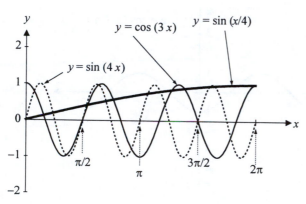

$y = \cos(3x)$ $y = \sin(x/4)$

$y = \sin(4x)$

$\pi/2$ π $3\pi/2$ 2π

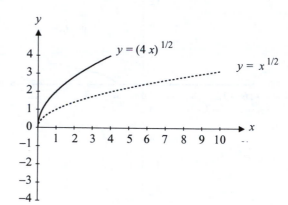

$y = (4x)^{1/2}$

$y = x^{1/2}$

3)

4)

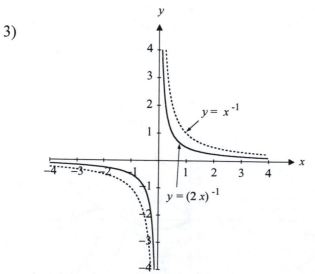

$y = x^{-1}$

$y = (2x)^{-1}$

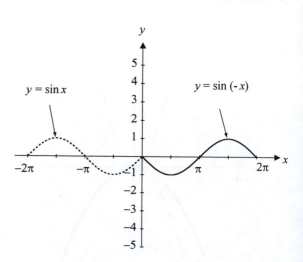

$y = \sin x$ $y = \sin(-x)$

3.8 page 48

1)
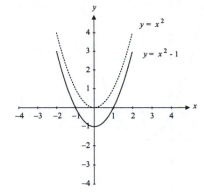

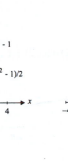
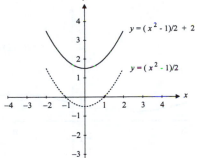

2)
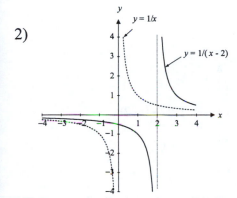
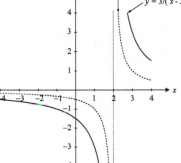
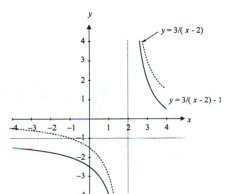

3) and 4)
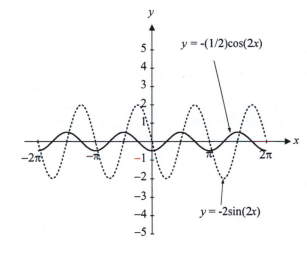

5)

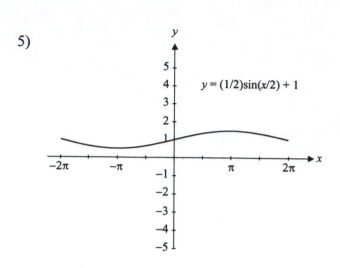

$y = (1/2)\sin(x/2) + 1$

6)

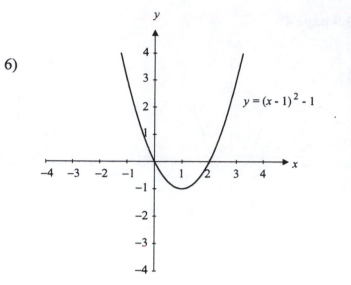

$y = (x - 1)^2 - 1$

7)

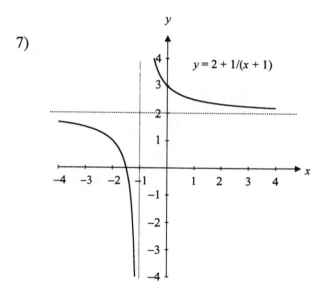

$y = 2 + 1/(x + 1)$

8)

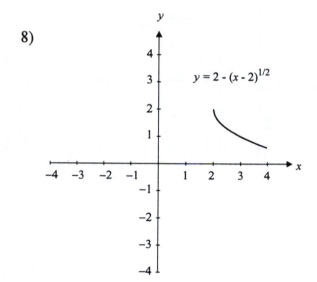

$y = 2 - (x - 2)^{1/2}$

9)

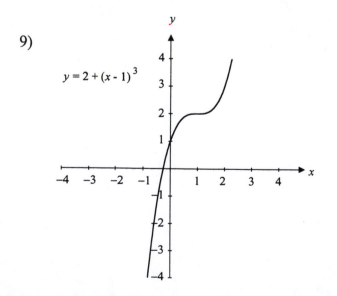

$y = 2 + (x - 1)^3$

10)

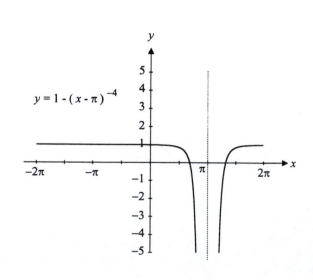

$y = 1 - (x - \pi)^{-4}$

Chapter 4

4.1 page 49

1) $2x(y+2)$ 2) $2wz(3+t)$ 3) $x(y+4+2w)$ 4) $3xy(2x+1+3y)$

5) $5x^2 y^4 (2x^6 y^2 + 5 + 4xy^6)$ 6) $\sin x \cos y (2\sin x + \cos y)$ 7) $2xyz(12x + yz + 2z^2)$

4.2 page 53

1) $(2y+3z)(2y-3z)$

2) $(4x^2 - y^3)(4x^2 + y^3)$

3) $(2s+3t)(4s^2 - 6st + 9t^2)$

4) $(2^{\frac{1}{3}}x + 4y)(2^{\frac{2}{3}}x^2 - 4\cdot 2^{\frac{1}{3}}xy + 16y^2)$

5) $(2s-3t)(4s^2 + 6st + 9t^2)$

6) $(4z - 9^{\frac{1}{3}}t)(16z^2 + 4\cdot 9^{\frac{1}{3}}zt + 9^{\frac{2}{3}}t^2)$

7) $(x+1)^2$

8) $(x+4)(x+2)$

9) $(x-6)(x+4)$

10) $(a - \sqrt{6})(a + \sqrt{6})(a^2 + 4)$

11) $(s-2)(s^2 + 2s + 4)(s+1)(s^2 - s + 1)$

12) $(3x-2)(x+1)$

4.3 page 55

1) $(3x + 2y)(a + b)$

2) $(x-y)(x+1)(x^2 - x + 1)$

3) $(x^4 + y^2)(x^2 + y)(x^4 - x^2 y + y^2)$

4) $(\sqrt{3}x + \sqrt{2}y)(\sqrt{3}x - \sqrt{2}y)(2y)(x+2)$

5) $(x+y)(3x + 5y + 7)$

4.4 page 59

1) $(x-1)^2(x+2)$ 2) $(x+1)(2x^2 - 2x + 3)$ 3) $(x+2)(x^2 - 2x + 3)$

4) $2x^2 - 3x + 4$ 5) $\left(x - \dfrac{3 + \sqrt{17}}{2}\right)\left(x - \dfrac{3 - \sqrt{17}}{2}\right)$ 6) $24(x-3)(x+1)$

4.5 page 62

1) $7(\sqrt{2} + 1)$ 2) $\dfrac{x^2 - 2}{2(x - \sqrt{2})}$ 3) $\dfrac{3(x + \sqrt{7})}{x^2 - 7}$ 4) $\dfrac{(x+1)(x - \sqrt{11})}{x^2 - 11}$

5) $x + \sqrt{3}$ 6) $(x^2 + 6)(x - \sqrt{6})$ 7) $(x^4 + 3)(x^2 - \sqrt{3})$

8) $\dfrac{-2}{\sqrt{2x+2h}\ \sqrt{2x}\left(\sqrt{2x} + \sqrt{2x+2h}\right)}$

4.6 page 65

1) $4|x|$ 2) $2|x|\sqrt{1 + 2x^2}$ 3) $3x\sqrt[3]{2x}$ 4) $x^6\sqrt{3y}$

5) $x^2\sqrt{5 + 3x^4}$ 6) $3x^2 \cos^{\frac{1}{3}} x$ 7) $2\pi y^2 x\sqrt{2x}$, for $x \geq 0$

8) $|xy|\sqrt[4]{x + x^2 y^6}$

Chapter 5

5.1 page 70

1) $\dfrac{1}{x(x-1)}$

2) $\dfrac{x^2 - 2x + 2}{x(x-1)}$

3) $\dfrac{2x}{(x^2 - 1)}$

4) $2(s^2 + 1)$

5) 2

6) $\dfrac{-2x - h}{x^2(x+h)^2}$

7) $\dfrac{x^4 + 3x^2 + 1}{x^3 + 2x}$

8) $\dfrac{2}{\sqrt{2x + 2h} + \sqrt{2x}}$

9) $\dfrac{a - 9b}{75a^5 b^7}$

10) $3x^2 + 2 + 3xh + h^2$

11) $\dfrac{1}{\sqrt{x + \Delta x - 3} + \sqrt{x - 3}}$

Chapter 6

6.1 page 74

1) a) $\dfrac{2\pi}{3}$
b) $\dfrac{3\pi}{2}$
c) $\dfrac{3\pi}{4}$
d) $\dfrac{7\pi}{6}$
e) $\dfrac{-5\pi}{6}$
f) $\dfrac{5\pi}{2}$

2) a) $135°$
b) $330°$
c) $-60°$
d) $540°$
e) $810°$
f) $405°$

6.2 page 78

1) a) -1 b) 0 c) 0 d) 0

2) a) 1 b) 0 c) 0 d) -1 if k is odd, and 1 if k is even

6.3 page 79

1) undefined, 1, undefined, 0

2) undefined, -1, undefined, 0

3)
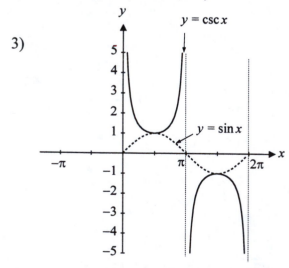

4)
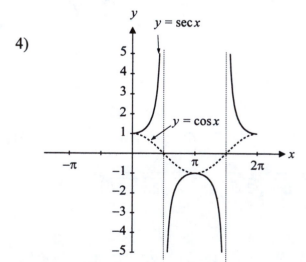

5)
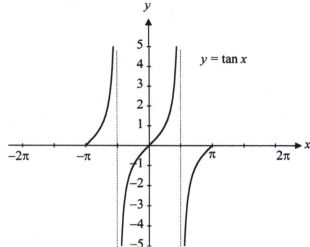

6)
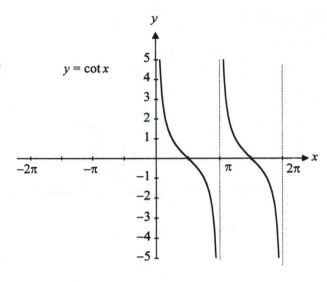

7)
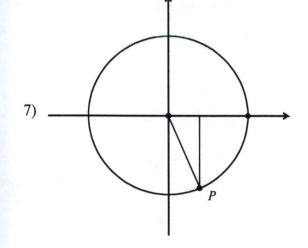

-1.2 is a little bigger than $-1.5 \cong \dfrac{-\pi}{2}$, and

so $\cos(-1.2) = $ 1st coordinate of $P \approx .4$.

8)
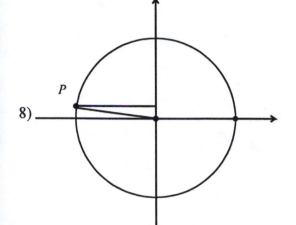

3.0 is a very small amount less than π, so
$\sin(3.0) = $ 2nd coordinate of $P \approx .1$.

6.4 page 81

1) a)

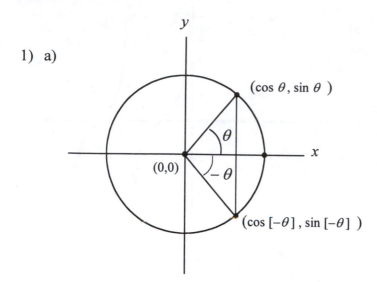

b) $\cos \theta = \cos (-\theta)$ c) $\sin (-\theta) = -\sin \theta$ d) $\tan (-\theta) = -\tan \theta$

2) a) $\tan \theta = \pm \sqrt{\sec^2 \theta - 1}$ b) $\cot \theta = \pm \sqrt{\csc^2 \theta - 1}$

6.5 page 86

1) $\dfrac{1}{\sqrt{2}}$ 2) -2 3) -2 4) $\sqrt{3}$ 5) $\dfrac{-1}{\sqrt{2}}$ 6) $\dfrac{-1}{\sqrt{2}}$ 7) -1

8) -2 9) $\dfrac{-2}{\sqrt{3}}$

Chapter 7

7.1 page 90

1) a) $\cos^3 x$ b) $(2t + 1)^3$ c) $(\sin x - 4x)^3$ d) $\cos x^3$
   e) $\cos (\cos x)$ f) $\cos (\sin x - 4x)$

2) a) $\tan x - 2\sqrt{\tan x}$ b) $\tan^2 \sqrt{x} - 2 \tan \sqrt{x}$ c) $\tan^{\frac{1}{2}} (x^2 - 2x)$
   d) $\sqrt{\tan^2 x - 2 \tan x}$ e) $(x - 2\sqrt{x})^2 - 2(x - 2\sqrt{x})$ f) $\sqrt[4]{x^2 - 2x}$

7.2 page 92

1) Outer function $f(x) = x^2$, inner function $g(x) = \tan x$.
2) Outer function $f(x) = x^2$, inner function $g(x) = x^3 - 1$.
3) Outer function $f(x) = \sin x$, inner function $g(x) = \sqrt{x}$.

4) Outer function $f(x) = \cos x$, inner function $g(x) = x^5$.

5) Outer function $f(x) = x^{2/3}$, inner function $g(x) = \sqrt[5]{x} - 1$.

6) $f(x) = \sin x$, $g(x) = \sqrt{x+1}$.

7) $f(x) = x^3$, $g(x) = \tan 2x$.

8) $f(x) = \cos x$, $g(x) = \left(x^3 - 2\right)^{3/4}$.

Chapter 8

8.1 page 97

1) 84   2) $\dfrac{75}{7}$ 3) $\dfrac{-3z - \pi}{2z}$ 4) $\dfrac{4z^2 y^2 - \pi}{2zy^2 + 2y}$ 5) $\dfrac{-2y - 3xy^3}{x + 3x^2 y^2}$

6) $\dfrac{-xy - 2xy^2 + 3x}{x + 2x^2 y + 2y}$ 7) $y' = \dfrac{-1 - \sin x}{\cos x - 2x}$

8.2 page 103

1) $-1, -4$ 2) $2 \pm i$ 3) -3 4) $\dfrac{-7 \pm \sqrt{33}}{2}$ 5) 4   6) ± 2

7) ± 8 8) $\pm 2\sqrt{2}, \pm 2\sqrt{2}\,i$ 9) $\pm \sqrt{5}, \pm 2$ 10) $3, -1$

11) $-y - z \pm \sqrt{3y^2 + 2yz - z^2}$

12) $\dfrac{-x \pm \sqrt{-x^2 - 4xy + 4y^2}}{2}$ 13) 4   14) $0, \pi, 2\pi, \dfrac{\pi}{2}$

15) $\cos x = \dfrac{1}{2}$ or -1, so $x = \pm \pi$ or $\pm \dfrac{\pi}{3}$

Chapter 9

9.1 page 109

1) Let L = the length of the fencing, and x = the width of the field in feet.
   Then $L = 2x + \dfrac{20{,}000}{x}$ feet.

2) Let P = the perimeter of the window, A = the area of the window, and x = the width of the window
   in feet. Then $P = 5x + \dfrac{\pi}{2}x$, and $A = x^2\left(2 + \pi/8\right)$ feet^2.

3) If we let r = the radius of the can and h = the height of the can, both in inches, then
   $h = \dfrac{100 - 2\pi r^2}{2\pi r}$ inches.

4) If we let $V = $ the volume of the box and $x = $ the length of the square cut out, both in inches, then
$V = 4x^3 - 36x^2 + 80x$ in^3.

5) With variables defined as in exercise (4) with x in meters we have $V = 4x^3 - 7x^2 + 3x$ m^3.

   If $A = $ the exterior surface area then $V = 4\left(\dfrac{3-A}{4}\right)^{\frac{3}{2}} - 7\left(\dfrac{3-A}{4}\right) + 3\left(\dfrac{3-A}{4}\right)^{\frac{1}{2}}$ m^3.

6) $V = \dfrac{S^{\frac{3}{2}}}{6\sqrt{\pi}}$ and $S = \sqrt[3]{36\pi} \, V^{\frac{2}{3}}$.   7) $F \cong 139.8$ lb.   8) 31%

9) $A = \dfrac{x^2}{4\pi} + \dfrac{(2-x)^2}{18}$

9.2 page 114

1) $z = \sqrt{34} \cong 5.8, \alpha \cong 31°$, and $\beta \cong 59°$.
2) $\beta = 50°$, $y \cong 11.9$, and $z \cong 15.6$.
3) $\beta = 58°$, $z \cong 9.4$, and $x \cong 5$.
4) $\alpha = 36°$, $y \cong 6.9$, and $z \cong 8.5$.
5) Angle at top $= 100°$, base $\cong 2.57h$, right side $= 2h$, , and left side $\cong 1.31h$.
6) The maximum vertical extent of the ladder is approximately 45.1 feet, so the highest it can reach is 49.1 feet. The floor level of the sixth floor is 45 feet, so the top of the ladder is roughly in the middle of the window. Perfect!
7) Rowing time is $\dfrac{3}{2}\sec\theta$, walking time is $\dfrac{1}{4}(10 - 3\tan\theta)$, so the total time is

   $\dfrac{1}{4}(6\sec\theta + 10 - 3\tan\theta)$.

9.3 page 118

1) $C = 30°$, $c \cong 7.83$, and $a \cong 15.43$.
2) $c \cong 5.57$, $A \cong 69°$, and $B \cong 51°$.
3) $A \cong 41.4°$, $B \cong 55.8°$, and $C \cong 82.8°$.
4) Case 1: $B \cong 58.8°$, $A \cong 101.2°$, and $a \cong 5.74$.
   Case 2: $B \cong 121.2°$, $A \cong 38.8°$, and $a \cong 3.66$.
5) The Pythagorean theorem.

Chapter 10

10.1 page 123

1) $\cos x (1 - \sin^2 x)^3$

2) $\pm \sec^2 x \left(\sqrt{1 + \tan^2 x} \right)^3$

3) $\sec x \tan x (\sec^2 x - 1)^2$

5) $\tan 2A = \dfrac{2 \tan A}{1 - \tan^2 A}$

7) $\cot 2A = \dfrac{\cot^2 A - 1}{2 \cot A}$

8) Use the result in question 6 with $A = \dfrac{\pi}{2}$, and $B = \theta$.

9) Recall that $\csc\left(\dfrac{\pi}{2} - \theta \right) = \dfrac{1}{\sin\left(\dfrac{\pi}{2} - \theta \right)} = \dfrac{1}{\sin\dfrac{\pi}{2} \cos\theta - \cos\dfrac{\pi}{2} \sin\theta} = \dfrac{1}{\cos\theta} = \sec\theta$

Appendix A

A.1 page 127

1) $y = (2/3)^x$ $y = .32^x$ $y = (3/2)^x$ $y = 1.1^x$

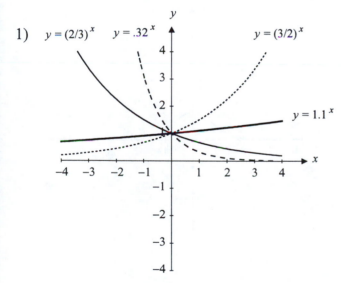

As $x \to \infty$, $.32^x$ and $(\tfrac{2}{3})^x$ both approach 0, while $(\tfrac{3}{2})^x$ and 1.1^x both approach ∞.

As $x \to -\infty$, $.32^x$ and $(\tfrac{2}{3})^x$ both approach ∞, while $(\tfrac{3}{2})^x$ and 1.1^x both approach 0.

2)

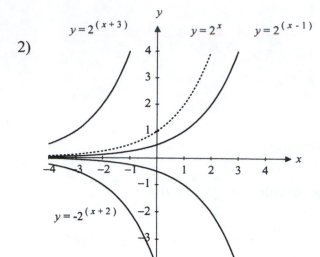

3)

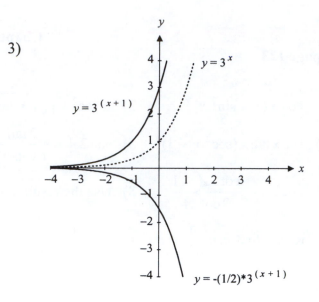

A.2 page 128

1)

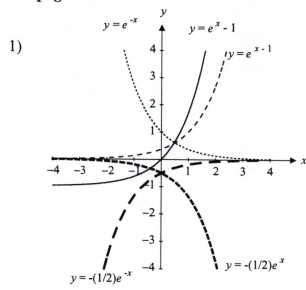

2)

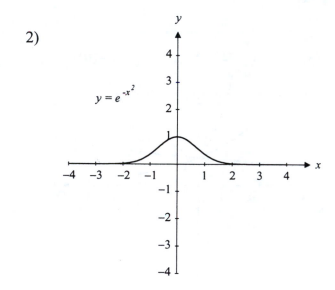

Appendix B

B.2 page 135

1)

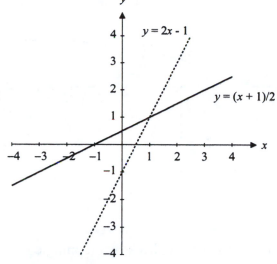

$y = 2x - 1$

$y = (x + 1)/2$

2)

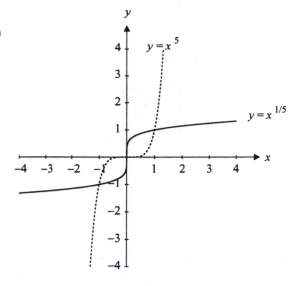

$y = x^5$

$y = x^{1/5}$

3) The function $g(x)$ is its own inverse.

4) An inverse does not exist, because $\cos y$ flunks the horizontal line test on the real line.

5)

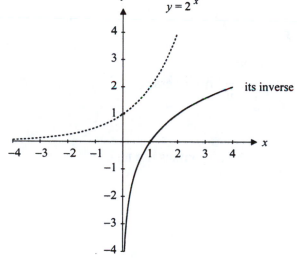

$y = 2^x$

its inverse

6) Also flunks the horizontal line test.

7)

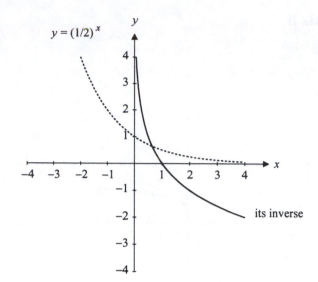

$y = (1/2)^x$

its inverse

8)

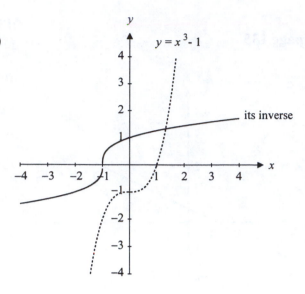

$y = x^3 - 1$

its inverse

9) On $[0, \pi]$ $\cos x$ has an inverse, but on $\left[-\dfrac{\pi}{2}, \dfrac{\pi}{2} \right]$ no inverse exists. (Check by graphing $\cos x$.)

B.3 page 138

1) $f^{-1}(x) = \dfrac{x + 3}{2}$

2) $k^{-1}(x) = \dfrac{x}{1 - x}$

3) $g^{-1}(x) = \dfrac{x^3 - 1}{5}$

4) $s^{-1}(t) = t^2 - 2$

5) $f^{-1}(x) = \dfrac{2}{x}$

6) $w = \pm \sqrt{\dfrac{v}{1 - v}}$, no inverse!

7) It flunks the horizontal line test; no inverse. However, if $x \geq 1$, it passes the test.

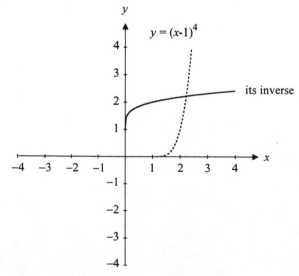

$y = (x-1)^4$

its inverse

8a) Domain is [0,1], and range is [3,4]. The inverse is the function $\sqrt{x-3} = (x-3)^{1/2}$.

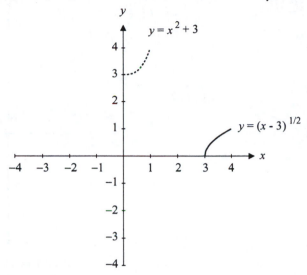

8b) Domain is [−1,0], and range is [3,4]. The inverse is the function $-\sqrt{x-3} = -(x-3)^{1/2}$.

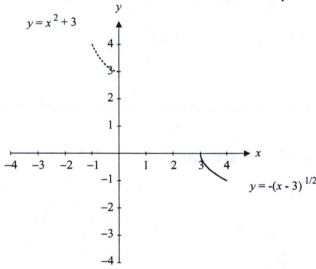

8c) No, their inverses are not the same, hence domains are very important when discussing inverse functions.

Appendix C

C.1 page 140

1) 2   2) − 1   3) −3   4) −3   5) $\dfrac{1}{2}$ 6) $\dfrac{3}{4}$ 7) 13   8) x

C.2 page 146

1)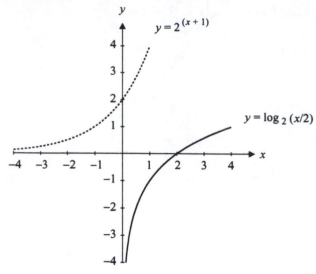

2)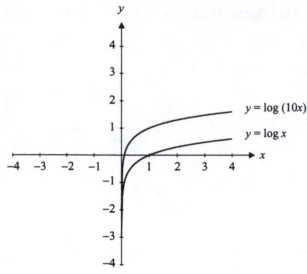

Notice that $\log 10x = 1 + \log x$.

3) a) 9   b) 2   c) $\dfrac{5}{2}$ d) -8

4) $\dfrac{-20}{3}$ 5) ± 2 6) $\pm\sqrt{14}$ 7) $0, 2\pi$

C.3 page 149

1) 9   2) $\left(\dfrac{1}{10}\right)^{\frac{1}{3}}$ 3) $\dfrac{7}{10}$ 4) 4,8

5) Let $a = 2, x = y = 1$. R.H.S. $= 0$, but L.H.S. $= 1$

6) 1.26

C.4 page 153

1)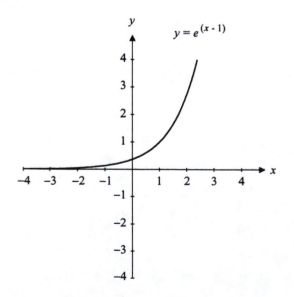

2)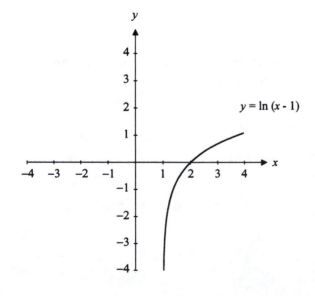

3)

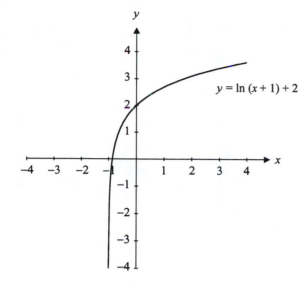

$$y = \ln(x+1) + 2$$

4) e^2 5) $1, -3$ 6) ± 2 7) $e^{2.3x},\ e^{-.69x}$

Appendix D

D.1 page 156

1) $x^6 + 6x^5 + 15x^4 + 20x^3 + 15x^2 + 6x + 1$

2) $y^6 - 6y^5 + 15y^4 - 20y^3 + 15y^2 - 6y + 1$

3) $16z^4 + 96z^3 + 216z^2 + 216z + 81$

4) $x^5 + 5x^4\Delta x + 10x^3(\Delta x)^2 + 10x^2(\Delta x)^3 + 5x(\Delta x)^4 + (\Delta x)^5$

5) $5x^4 + 10x^3(\Delta x) + 10x^2(\Delta x)^2 + 5x(\Delta x)^3 + (\Delta x)^4$

Index